SpringerBriefs in Mathematics

SpringerBriefs in Mathematics showcases expositions in all areas of mathematics and applied mathematics. Manuscripts presenting new results or a single new result in a classical field, new field, or an emerging topic, applications, or bridges between new results and already published works, are encouraged. The series is intended for mathematicians and applied mathematicians.

For further volumes:
http://www.springer.com/series/10030

Wei Zeng • Xianfeng David Gu

Ricci Flow for Shape Analysis and Surface Registration

Theories, Algorithms and Applications

 Springer

Wei Zeng
Computing and Information Science
Florida International University
Miami, FL, USA

Xianfeng David Gu
Computer Science
State University of New York
Stony Brook, NY, USA

ISSN 2191-8198　　　　　　　ISSN 2191-8201 (electronic)
ISBN 978-1-4614-8780-7　　　ISBN 978-1-4614-8781-4 (eBook)
DOI 10.1007/978-1-4614-8781-4
Springer New York Heidelberg Dordrecht London

Library of Congress Control Number: 213946911

Mathematics Subject Classification (2010): 52, 52-01, 52-02

Printed on acid-free paper

Springer is part of Springer Science+Business Media (www.springer.com)

Preface

Ricci flow deforms the Riemannian metric proportionally to the curvature, such that the curvature evolves according to a heat diffusion process and eventually becomes constant everywhere. Ricci flow is a powerful tool in geometric analysis for studying low dimensional topology. It has been successfully applied for the proofs of Poincaré's conjecture and Thurston's geometrization conjecture. Recently, Ricci flow has started making impacts on practical fields and tackling fundamental engineering problems. This book focuses on the theories and algorithms of discrete surface Ricci flow, and its applications on surface registration and shape analysis.

General Ricci flow is defined on arbitrary dimensional Riemannian manifolds. Surface (two-manifold) Ricci flow has unique characteristics, which are crucial for developing discrete theories and designing computational algorithms. *First*, surface Ricci flow never blows up, namely, the Gauss curvature during the flow is always bounded. This phenomenon ensures the numerical stability of discrete surface Ricci flow. In contrast, three-manifold Ricci flow will produce singularities; thus topological surgery is unavoidable. *Second*, surface Ricci flow is conformal, namely, the deformation of the Riemannian metric preserves angles. This fact greatly simplifies both theoretical arguments and algorithmic designs. General Ricci flow is governed by tensor differential equations, whereas surface Ricci flow is described by scalar differential equations. *Third*, surface Ricci flow has intuitive geometric interpretations, which directly lead to the design of data structures. A conformal deformation transforms infinitesimal circles to infinitesimal circles. This elucidates the geometric nature of the flow. *Finally*, Ricci flow is variational, namely, Ricci flow is the negative gradient flow of Ricci energy. Accordingly, discrete surface Ricci flow can be formulated as a convex optimization problem, which has a unique global optimum and can be carried out using the efficient Newton's method.

For the purpose of surface registration and shape analysis, discrete surface Ricci flow has the following unique merits: (a) by Ricci flow, all shapes in real life can be unified to one of the following three canonical shapes: the sphere, the plane, or the hyperbolic disk; (b) therefore, most 3D geometric problems can be converted to 2D image problems, which greatly simplifies the computation; (c) furthermore, this conversion is conformal and preserves the original geometric information; (d) finally, by deforming Riemannian metric, Ricci flow can be used to compute general diffeomorphisms between surfaces.

Ricci flow has demonstrated its great potential by solving various problems in many fields, which can be hardly handled by alternative methods so far. The following are some examples: (1) nonrigid surface registration and tracking in computer vision, (2) global surface parameterization in computer graphics, (3) conformal brain mapping and virtual colonoscopy in medical imaging, (4) the shortest word problem in computational topology, (5) delivery guaranteed greedy routing and load balancing in wireless sensor network, and so on. We believe that more and more researchers will realize and appreciate the intrinsic power and beauty of Ricci flow, and more and more fields in engineering and medicine will be impacted by Ricci flow.

This book is mainly for graduate students and researchers in the fields of computer science, applied mathematics, engineering, and medical imaging. The book provides both theoretical foundations and computational methods for surface Ricci flow. The introduction to the smooth geometry theories is self-contained. The discrete theories and computational algorithms are written using elementary mathematical tools, and all the details are well exposed, such that students with engineering background can easily follow and digest them. In order to help students and researchers reproduce the algorithms in the book for their own research projects, sample codes and data sets are available on the authors' web sites: http://www.cs. stonybrook.edu/~gu and http://www.cs.fiu.edu/~wzeng. These computational tools are also valuable for professionals in the fields related to surface registration and shape analysis, such as digital media and digital entertainment industry, geometric modeling and computer aided design industry, medical imaging industry, and biometrics industry.

In this book, the first chapter (Chap. 1) gives an overview of the whole contents; the rest of the book is organized into two parts. The first part (Chaps. 2 and 3) gives brief introduction to the theoretical foundations necessary to understand Ricci flow, mainly algebraic topology, surface differential geometry, and Riemann surface theory; the second part (Chaps. 4 and 5) provides complete proofs, algorithmic details for discrete surface Ricci flow, and applications in practice. Students emphasizing engineering applications may start reading the second part directly. An overview of each chapter in the book is as follows:

- Chapter 1 introduces the fundamental concepts of shape space and mapping space, including different transformation groups (such as diffeomorphisms, isometries, conformal transformations, and rigid motions) and group actions on shape spaces. In order to perform surface registration and shape analysis in the shape space and the mapping space, Ricci flow is introduced, which leads to the celebrated uniformization theorem.
- Chapter 2 briefly reviews the fundamental concepts and theorems in algebraic topology, surface differential geometry, and surface Ricci flow.

- Chapter 3 briefly introduces the Riemann surface theory, including quasi-conformal mapping, Teichmüller space, and surface harmonic maps. Finally, the Teichmüller theory of harmonic maps is covered.
- Chapter 4 systematically introduces the discrete surface Ricci flow theory. The whole theory is explained thoroughly using variational principle on discrete surfaces based on derivative cosine law. Complete proofs for most theorems and lemmas are given in detail.
- Chapter 5 focuses on the computational algorithms and direct application examples. The algorithms have been fully tested to handle various problems in real world for many years and are mature for broad practical applications.

This book summarizes the research of our group in the past 10 years. We would like to thank all our collaborators, especially, Shing-Tung Yau, Feng Luo, Huaidong Cao, David Mumford, Stanley Osher, David Glickenstein, Paul Thompson, Tony F. Chan, Zhonglin Lu, Jun Zhang, Warner Miller, Harry Shum, Arie E. Kaufman, Sundaraja S. Iyengar, Hong Qin, Dimitris Samaras, Jie Gao, Jerome Liang, Song Zhang, Yalin Wang, Shi-Min Hu, Zhongxuan Luo, Ronald Lok Ming Lui, Ren Guo, Jian Sun, and many others. We are indebted to all the group members, Miao Jin, Ying He, Junho Kim, Xiaotian Yin, Xin Li, Ruirui Jiang, Yinghua Li, Min Zhang, Mayank Goswami, Xiaokang Yu, Rui Shi, Zhengyu Su, and so on. We are very grateful to the National Science Foundation (NSF), the Air Force Office of Scientific Research (AFOSR), the Office of Naval Research (ONR), and the National Institutes of Health (NIH) for their persistent support. We are also grateful to the Springer editors, Eve Mayer and Vaishali Damle, and the editorial director, Marc Strauss, for their great help during the book creation.

<table>
<tr><td>Miami, FL, USA</td><td>Wei Zeng</td></tr>
<tr><td>Stony Brook, NY, USA</td><td>Xianfeng David Gu</td></tr>
</table>

Contents

Chapter 1
Introduction

This chapter briefly introduces the fundamental concepts of shape space and mapping space, including different transformation groups (such as diffeomorphisms, isometries, conformal transformations, and rigid motions) and group actions on shape spaces. In order to perform surface registration and shape analysis in the shape space and the mapping space, Ricci flow is introduced, which leads to the celebrated uniformization theorem.

1.1 Manifold and Riemannian Metric

Definition 1.1 (Manifold). Let M be a topological space, and $\{U_\alpha\}$, $\alpha \in I$ be an open covering of M, $M \subset \cup_\alpha U_\alpha$. For each U_α, $\phi_\alpha : U_\alpha \to \mathbb{R}^n$ is a homeomorphism. The pair (U_α, ϕ_α) is called a chart. Suppose $U_\alpha \cap U_\beta \neq \emptyset$. The transition function $\phi_{\alpha\beta} : \phi_\alpha(U_\alpha \cap U_\beta) \to \phi_\beta(U_\alpha \cap U_\beta)$ is defined as

$$\phi_{\alpha\beta} = \phi_\beta \circ \phi_\alpha^{-1},$$

then M is called a manifold, $\{(U_\alpha, \phi_\alpha)\}$ is called an atlas, as shown in Fig. 1.1. If all transition functions $\phi_{\alpha\beta}$ are smooth, then M is a smooth manifold.

Two-dimensional manifolds are called *surfaces*. We can assign a Riemannian metric to the manifold.

Definition 1.2 (Riemannian Metric). Suppose for every point p in a manifold M, an inner product $\langle \cdot, \cdot \rangle_p$ is defined on a tangent space of M at p, T_pM. Then the collection of all these inner products is called the Riemannian metric.

A manifold with a Riemannian metric is called a *Riemannian manifold* and denoted as $(M, \mathbf{g})$, where $\mathbf{g}$ is the metric tensor. On a local chart (U_α, ϕ_α) with parameters (x_α, y_α), the metric tensor is represented as a symmetric positive definite matrix,

W. Zeng and X.D. Gu, *Ricci Flow for Shape Analysis and Surface Registration*,
SpringerBriefs in Mathematics, DOI 10.1007/978-1-4614-8781-4_1,
© Wei Zeng, Xianfeng David Gu 2013

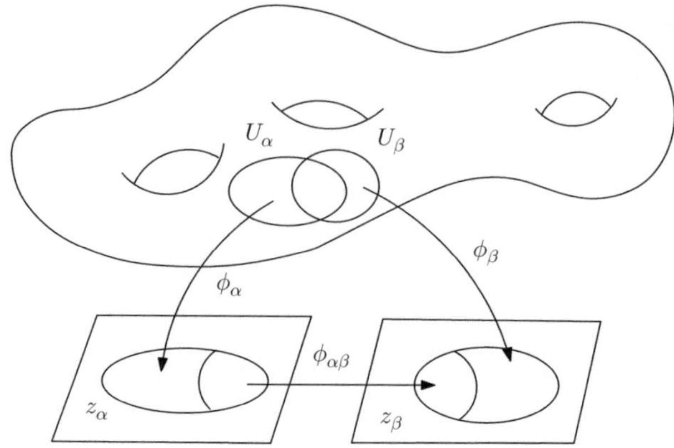

Fig. 1.1 A two-dimensional manifold with local charts

$$\mathbf{g} = \begin{pmatrix} g_{11}^{\alpha} & g_{12}^{\alpha} \\ g_{21}^{\alpha} & g_{22}^{\alpha} \end{pmatrix}.$$

Suppose $(U_{\beta}, \phi_{\beta})$ is another chart with local parameters (x_{β}, y_{β}). The Jacobian matrix for the coordinate change is

$$\frac{\partial(x_{\beta}, y_{\beta})}{\partial(x_{\alpha}, y_{\alpha})} = \begin{pmatrix} \frac{\partial x_{\beta}}{\partial x_{\alpha}} & \frac{\partial x_{\beta}}{\partial y_{\alpha}} \\ \frac{\partial y_{\beta}}{\partial x_{\alpha}} & \frac{\partial y_{\beta}}{\partial y_{\alpha}} \end{pmatrix},$$

then

$$\mathbf{g} = \begin{pmatrix} g_{11}^{\alpha} & g_{12}^{\alpha} \\ g_{21}^{\alpha} & g_{22}^{\alpha} \end{pmatrix} = \left[\frac{\partial(x_{\beta}, y_{\beta})}{\partial(x_{\alpha}, y_{\alpha})} \right]^{T} \begin{pmatrix} g_{11}^{\beta} & g_{12}^{\beta} \\ g_{21}^{\beta} & g_{22}^{\beta} \end{pmatrix} \left[\frac{\partial(x_{\beta}, y_{\beta})}{\partial(x_{\alpha}, y_{\alpha})} \right].$$

The Riemannian metric can be used to measure length, angle, and area. Suppose $\mathbf{r}(x, y)$ is the position vector of the surface. A curve $C : [0, 1] \to M$ on the surface has a parametric representation $C(t) = \mathbf{r}(x(t), y(t))$. Then its length is given by

$$s = \int_{0}^{1} \frac{ds}{dt} = \int_{0}^{1} \sqrt{g_{11}\left(\frac{dx}{dt}\right)^{2} + 2g_{12}\frac{dx}{dt}\frac{dy}{dt} + g_{22}\left(\frac{dy}{dt}\right)^{2}}\, dt. \tag{1.1}$$

The area element is given by

$$dA = \sqrt{g_{11}g_{22} - g_{12}^{2}}\, dx\, dy. \tag{1.2}$$

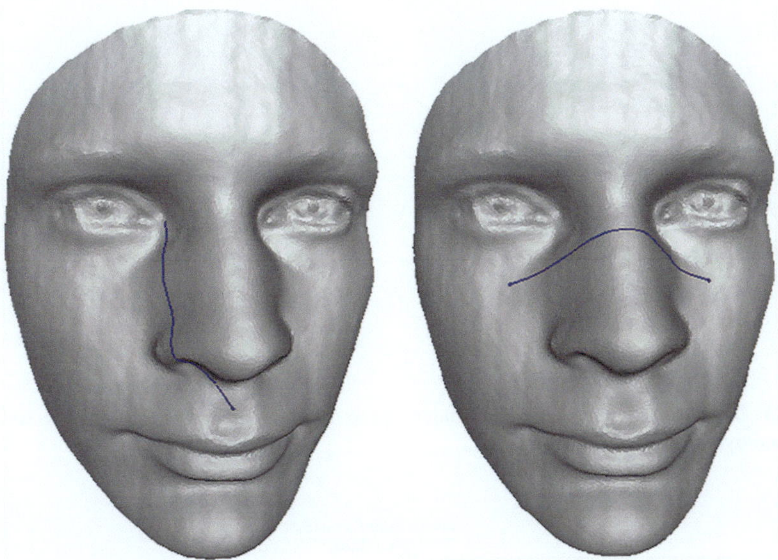

Fig. 1.2 Geodesics on a human face

Let $\Omega_0 \subset \Omega$ be a subdomain of Ω, then the area of the subsurface defined on Ω_0 is given by

$$\int_{\Omega_0} dA = \int_{\Omega_0} \sqrt{g_{11}g_{22} - g_{12}^2}\,dx\,dy.$$

Suppose two tangent vectors $\mathbf{v}_1 = a\mathbf{r}_x + b\mathbf{r}_y$ and $\mathbf{v}_2 = \alpha\mathbf{r}_x + \beta\mathbf{r}_y$ are at the same point p of the surface, $\mathbf{v}_1, \mathbf{v}_2 \in T_p M$ and the angle between them is θ. Then

$$\cos\theta = \frac{\langle \mathbf{v}_1, \mathbf{v}_2 \rangle}{|\mathbf{v}_1| \cdot |\mathbf{v}_2|}$$
$$= \frac{g_{11}a\alpha + g_{12}a\beta + g_{21}\alpha b + g_{22}b\beta}{\sqrt{g_{11}a^2 + 2g_{12}ab + g_{22}b^2}\sqrt{g_{11}\alpha^2 + 2g_{12}\alpha\beta + g_{22}\beta^2}}. \tag{1.3}$$

If two points on the manifold are selected, then there is a family of paths connecting them. *Geodesics* are those paths which have the extremal length. If the two points are close enough, then there exists a unique geodesic, which is the shortest, as shown in Fig. 1.2. Consider a geodesic triangle on a surface, whose three edges are geodesics, then the sum of the inner angles may not be equal to π, and the deviation of the sum from π is the *total Gauss curvature* inside the triangle. In fact, there are three canonical models for general surfaces, as shown in Fig. 1.3. The flat space model is the common Euclidean plane $\mathbb{E}^2$, where the curvature is zero everywhere; the unit sphere $\mathbb{S}^2$, where the curvature is $+1$ everywhere; and the hyperbolic plane $\mathbb{H}^2$, where the curvature is -1 everywhere. The cosine laws on different spaces are drastically different as follows,

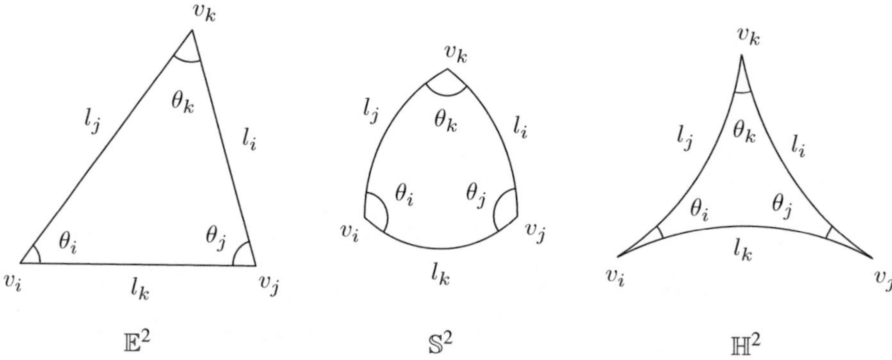

Fig. 1.3 Geodesic triangles on three canonical space models with constant curvatures, which are governed by different cosine laws

$$
\begin{aligned}
1 &= \frac{\cos\theta_i + \cos\theta_j \cos\theta_k}{\sin\theta_j \sin\theta_k} & \mathbb{E}^2 \\[4pt]
\cos l_i &= \frac{\cos\theta_i + \cos\theta_j \cos\theta_k}{\sin\theta_j \sin\theta_k} & \mathbb{S}^2 \\[4pt]
\cosh l_i &= \frac{\cosh\theta_i + \cosh\theta_j \cosh\theta_k}{\sinh\theta_j \sinh\theta_k} & \mathbb{H}^2
\end{aligned}
\tag{1.4}
$$

1.2   Ricci Flow

Curvatures are determined by Riemannian metrics. One natural question to ask is whether the metric can be determined by the curvature. In practice, it is highly desirable to design Riemannian metrics with the user prescribed curvatures. Hamilton's Ricci flow is a powerful tool to achieve such a goal. On surfaces, Hamilton defined the Ricci flow as

$$
\frac{dg_{ij}(t)}{dt} = -2K(t)g_{ij}(t),
$$

where $g_{ij}(t)$ and $K(t)$ are functions of time t, and $K(t)$ is the Gauss curvature induced by $g_{ij}(t)$. Basically, Ricci flow deforms the Riemannian metric proportionally to the curvature, such that the curvature evolves according to a nonlinear heat diffusion process, and eventually becomes constant everywhere. The curvature flow is represented as

$$
\frac{dK(t)}{dt} = \Delta_{g(t)}K(t) + 2K^2(t),
$$

where $\Delta_{g(t)}$ is the Laplace–Beltrami operator induced by the metric $g(t)$. Hamilton and Chow proved that during the flow, the curvature $K(t)$ is always finite, never

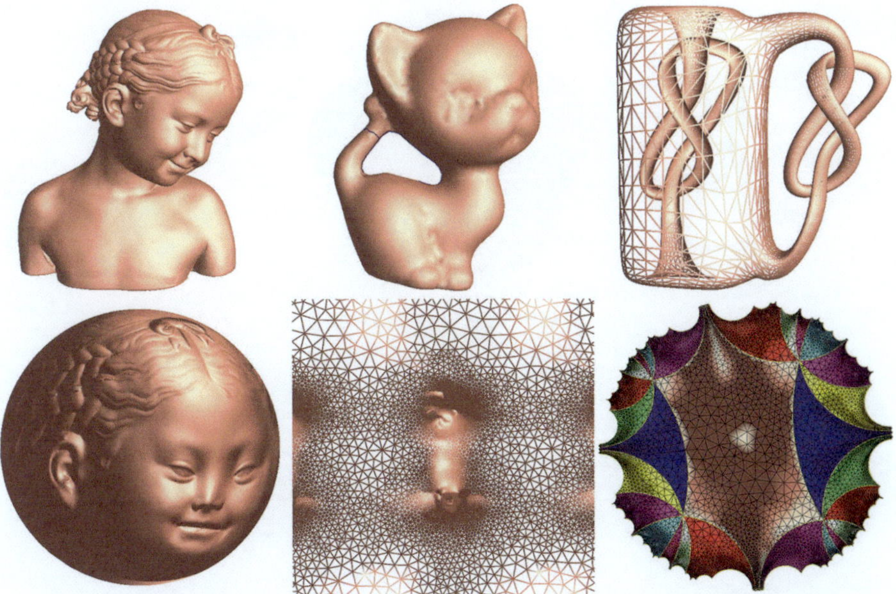

Fig. 1.4 Uniformization for closed surfaces

blows up, and when time goes to infinity, the curvature converges to constant, $K(\infty) \to const$. This leads to the celebrated *uniformization theorem*, which says all surfaces in real life can be deformed to one of the three canonical spaces, the sphere $\mathbb{S}^2$, the plane $\mathbb{E}^2$, or the hyperbolic disk $\mathbb{H}^2$. Figure 1.4 demonstrates the uniformization for closed surfaces. Surfaces with boundaries can be uniformized to the canonical spaces with circular boundaries. Figure 1.5 demonstrates the uniformization for surfaces with boundaries.

The uniformization transforms all the shapes in real life to canonical ones and converts all 3D surface geometric processing problems to 2D planar problems. This greatly simplifies most of the computational tasks and improves the efficiency and efficacy. The mappings among all surfaces with the same topology can be easily established by composing the mappings from the surfaces to the canonical spaces (uniformization transformations) and the automorphisms of the canonical spaces. This allows different shapes to be matched, registered, tracked, and compared.

1.3 Mappings Among Manifolds

Suppose M_1 and M_2 are two manifolds, with local charts (U_α, ϕ_α) and (V_β, ψ_β), respectively. A mapping $f : M_1 \to M_2$ has the local representation

$$\psi_\beta \circ f \circ \phi_\alpha^{-1} : \phi_\alpha(U_\alpha) \to \psi_\beta(V_\beta).$$

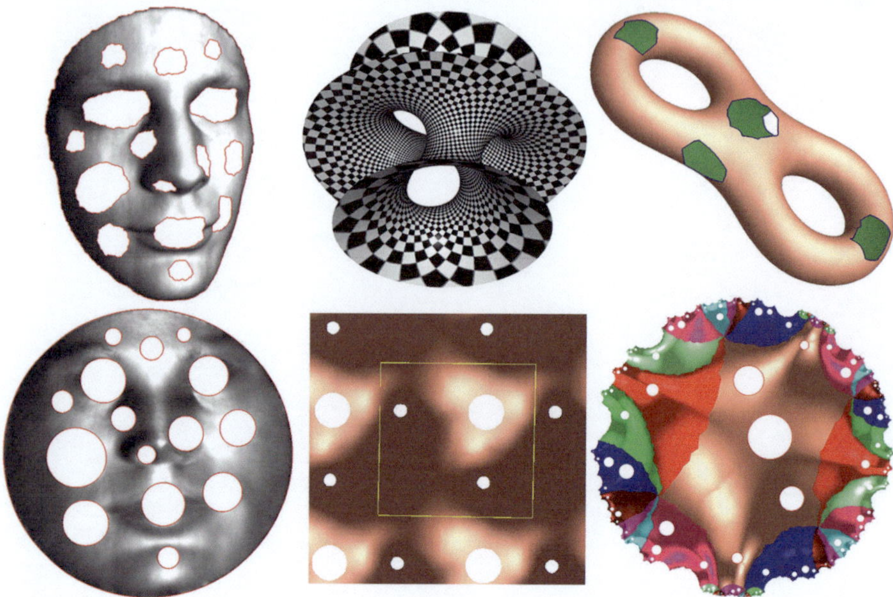

Fig. 1.5 Uniformization for surfaces with boundaries

For the convenience, we denote $\psi_\beta \circ f \circ \phi_\alpha^{-1}$ as f_α^β. Assume the local coordinates on $\phi_\alpha(U_\alpha)$ are (x, y), on $\psi_\beta(V_\beta)$ are (u, v), then $(u, v) := f_\alpha^\beta(x, y)$.

Homeomorphism

If all the local representations f_α^β are homeomorphisms, namely, they are continuous and invertible and their inverses are also continuous, then f is called a *homeomorphism* between the two manifolds, and we say M_1 and M_2 are *topologically equivalent*.

Diffeomorphism

Similarly, if all the local representations f_α^β are orientation preserving diffeomorphisms, namely, the determinant of the Jacobian matrix is positive everywhere,

$$\left| \frac{\partial(u, v)}{\partial(x, y)} \right| := \begin{vmatrix} \frac{\partial u}{\partial x} & \frac{\partial u}{\partial y} \\ \frac{\partial v}{\partial x} & \frac{\partial v}{\partial y} \end{vmatrix} > 0,$$

then we say f is a *diffeomorphism*, and the two manifolds are *diffeomorphically equivalent*.

Suppose the Riemannian metric on M_1 is $\mathbf{g}_1 = g_{11}^1 dx^2 + 2g_{12}^1 dxdy + g_{22}^1 dy^2$, and the Riemannian metric on M_2 is $\mathbf{g}_2 = g_{11}^2 du^2 + 2g_{12}^2 dudv + g_{22}^2 dv^2$. By a diffeomorphism f, a curve $C(t)$ on the source M_1 is mapped to a curve $f \circ C(t)$ on the target M_2, and its length can be measured by the metric $\mathbf{g}_2$. We can define the length of $C(t) \subset M_1$ as that of $f \circ C(t)$ on M_2. This gives another Riemannian metric on M_1, which is called the *pullback metric* induced by f, and denoted as $f^* \mathbf{g}_2$.

Definition 1.3 (Pullback Metric). Suppose $f : (M_1, \mathbf{g}_1) \to (M_2, \mathbf{g}_2)$ be a mapping between two Riemannian manifolds. The pullback metric induced by f has the local representation

$$f^* \mathbf{g}_2 = \left[\frac{\partial(u, v)}{\partial(x, y)} \right]^T \mathbf{g}_2 \left[\frac{\partial(u, v)}{\partial(x, y)} \right].$$

Isometric Mapping

Suppose the mapping f is a diffeomorphism, and the pullback metric induced by f equals the original metric

$$f^* \mathbf{g}_2 = \mathbf{g}_1.$$

Then the mapping is called an *isometry*. Isometric mappings preserve lengths.

Conformal Mapping

Suppose the pullback metric induced by a diffeomorphism f satisfies

$$f^* \mathbf{g}_2 = e^{2\lambda} \mathbf{g}_1,$$

where $\lambda : M_1 \to \mathbb{R}$ is a function defined on M_1. Then the mapping f is called a *conformal mapping* and $e^{2\lambda}$ is called the *conformal factor*. From (1.3), we can conclude that a conformal mapping preserves angles, therefore it is also called an *angle preserving mapping*.

Area Preserving Mapping

Suppose the pullback metric satisfies the following condition,

$$det(f^* \mathbf{g}_2) = det(\mathbf{g}_1).$$

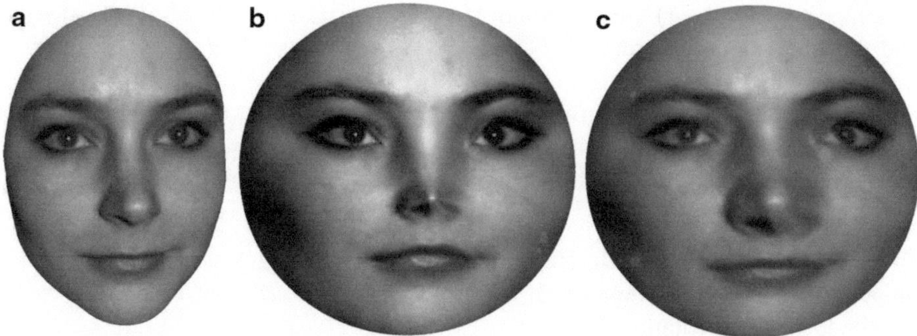

Fig. 1.6 Diffeomorphisms. (**a**) A 3D human face surface, (**b**) An angle preserving (conformal) mapping, (**c**) An area preserving mapping

Then the mapping is called an *area preserving mapping*, which preserves area element.

It is obvious that isometry is both angle preserving (conformal) and area preserving. The inverse is also true, i.e., if a mapping is both conformal and area preserving, then it must be isometric.

Figure 1.6 demonstrates angle preserving and area preserving mappings from a human face surface onto the planar disk. Frame **a** is the original facial surface, captured by a 3D scanner. Frame **b** is the image of a conformal mapping. Conformal mapping can be interpreted as local scaling transformations, so local shapes are well preserved. Frame **c** shows the area preserving mapping, where the area element is preserved.

Rigid Motion

Suppose both M_1 and M_2 are embedded in the Euclidean space $\mathbb{E}^n$, f is a rotation composed with a translation in $\mathbb{E}^n$. Then f is called a *rigid motion*.

In each category discussed above, all the mappings form a transformation group. For example, consider all the angle preserving mappings, if f, g are two conformal mappings, then their composition $g \circ f$ is still conformal; if f is conformal, then its inverse f^{-1} is conformal; the identity map is also conformal. Hence all conformal mappings form a transformation group.

1.4   Shape Space

We consider the space of all possible shapes. Here, shapes may refer to all one dimensional contours on the plane, or all surfaces embedded in three-dimensional Euclidean space. Assume their volumes are finite and compact. We denote the shape space as $\mathfrak{M}$, for example,

$$\mathfrak{M} = \{S \hookrightarrow \mathbb{E}^3\},$$

where S is a compact and orientable surface, embedded in $\mathbb{E}^3$. Let G be a transformation group that acts on the shape space $\mathfrak{M}$, such as conformal transformations,

$$G \times \mathfrak{M} \to \mathfrak{M}.$$

Then $g(S)$ is another surface, denoted as a pair $(g, S) \in \mathfrak{M}$. The orbits of G in $\mathfrak{M}$ can be defined as *equivalence classes*,

$$[S] = \{(g, S) | g \in G\}.$$

For example, if G is the conformal transformation group, then each conformal equivalence class $[S]$ is called a *Riemann surface*.

The quotient space $\mathfrak{M}/G$ is the set of such equivalence classes,

$$\mathfrak{M}/G = \{[S] | S \in \mathfrak{M}\}.$$

The space of all Riemann surfaces (with the same topology) is called the *moduli space*. The topology of moduli space is complicated. Instead, we study its universal covering space, the so-called *Teichmüller space*. For genus $g > 1$ closed surfaces, the Teichmüller space is a $6g - 6$ manifold, homeomorphic to $\mathbb{R}^{6g-6}$. We can design Riemannian metric for the quotient space $\mathfrak{M}/G$ and measure the distances among orbits. The distance between two Riemann surfaces in Teichmüller space is given by the so-called *Teichmüller map*, which is unique and minimizes the angle distortion. The distance is defined as the logarithm of the dilatation of the Teichmüller map, where the dilation is a measurement of angle distortions. Therefore, Teichmüller spaces are Riemannian manifolds.

In general, if $\mathfrak{M}$ has a Riemannian metric, the action of G on $\mathfrak{M}$ is isometric, namely, $g \in G$, $S_1, S_2 \in \mathfrak{M}$,

$$d_{\mathfrak{M}}(S_1, S_2) = d_{\mathfrak{M}}(g(S_1), g(S_2)),$$

then the geodesic distance in the quotient space $\mathfrak{M}/G$ is given by

$$d_{\mathfrak{M}/G}([S_1], [S_2]) = \min_{g \in G} d_{\mathfrak{M}}(S_1, g(S_2)).$$

1.5 Mapping Space

For practical purposes, we are only interested in diffeomorphisms (generally, approximated by homeomorphisms). The diffeomorphic mappings between two surfaces also form a space, which we call *mapping space*. The mapping space is

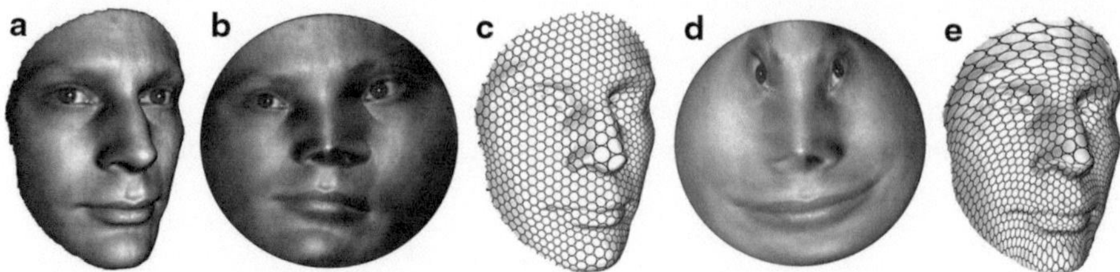

Fig. 1.7 Diffeomorphisms and Beltrami coefficients. (**a**) A 3D human face surface, (**b**) An angle preserving (conformal) mapping ($\mu = 0$), (**c**) Circle-packing texture mapping induced by the angle preserving mapping, (**d**) A quasi-conformal mapping ($0 < \|\mu\|_\infty < 1$), (**e**) Circle-packing texture mapping induced by the quasi-conformal mapping

Fig. 1.8 Geometric illustration of Beltrami coefficient. μ is the Beltrami coefficient, D is the eccentricity, which is the ratio between the major axis and the minor axis, and θ is the angle between the major axis and the horizontal direction

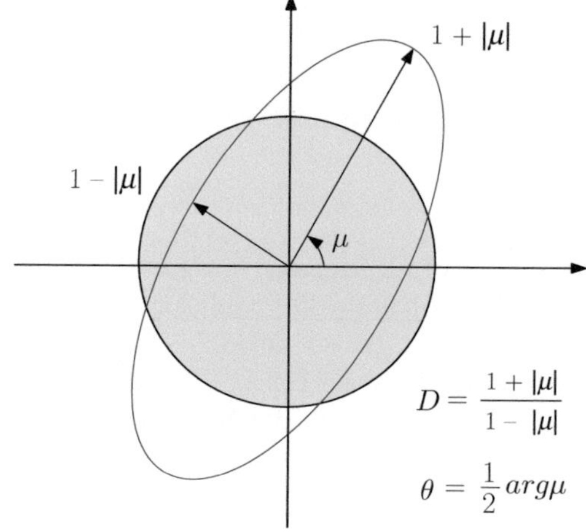

infinite dimensional. Given the source surface S_1 and the target surface S_2, all the mappings can be classified by their homotopy types. Fixing the homotopy class, each diffeomorphism $f : S_1 \to S_2$ corresponds to a unique complex function (or complex differential) μ_f, $\|\mu_f\|_\infty < 1$, called the *Beltrami coefficient* (or *Beltrami differential*) of the mapping f, where the L^∞ norm $\|\mu_f\|_\infty$ is defined as the maximum of the norm of μ on the source surface.

Figure 1.7 shows diffeomorphisms with different Beltrami coefficients. A quasi-conformal mapping maps infinitesimal circles on the source surface to infinitesimal ellipses on the target. The eccentricity of the ellipse at point p and the orientation are encoded to a complex number $\mu(p)$. Figure 1.8 explains the geometric meaning of Beltrami coefficient. Note that the size of the ellipse is not encoded; therefore, the Beltrami coefficient has less information than the Jacobian matrix of the mapping.

Amazingly, diffeomorphism can be fully recovered from its Beltrami coefficient. Essentially, each Beltrami coefficient μ uniquely determines a diffeomorphism. This converts the mapping space to a complex functional space $\{\mu | \mu : S_1 \to \mathbb{C}, \|\mu\|_\infty < 1\}$. Furthermore, the diffeomorphism f^μ depends on μ smoothly. The variation of the mapping with respect to the variation of its Beltrami coefficient has an explicit analytic relation. This allows us to perform optimization in the mapping space.

In practice, several special types of mappings are commonly used: (1) harmonic mappings, which minimize the membrane energy; (2) biharmonic mappings, which minimize the elastic energy; (3) conformal mappings, which preserve angles; (4) extremal quasi-conformal mappings, which minimize angle distortions; and (5) area preserving mappings, which preserve area elements.

1.6 Computational Frameworks

We briefly introduce the computational frameworks for surface classification, comparison, and registration based on Ricci flow. Details will be explained in later chapters.

1.6.1 Surface Classification

Surfaces are classified by different transformation groups. One transformation group G acts on the shape space $\mathfrak{M}$ and classifies the shape space to orbits. The orbits form the quotient space $\mathfrak{M}/G$. Different transformation groups correspond to different geometries and require different theoretical tools and computational methodologies.

Homeomorphism Group

The quotient space $\mathfrak{M}/G$ is a discrete point set. Two surfaces are in the same topological equivalence class if and only if they have the same genus g and the same boundary components b.

In practice, the surfaces are represented by polyhedron surfaces. The Euler characteristic number is given by

$$\chi(S) := 2 - 2g - b,$$

which can be computed by $\chi(S) = V + F - E$, where V, E, F are the number of vertices, edges, and faces of the polyhedron surface. The number of boundary components b can be calculated by tracing the boundary edges, then the genus g can be obtained.

The computational algorithms are designed mainly based on algebraic topology.

Conformal Transformation Group

The quotient space $\mathfrak{M}/G$ is a finite dimensional space, which is the Teichmüller space. Two surfaces are conformally equivalent if and only if they share the same conformal module.

By using Ricci flow, we can compute the uniformizations of surfaces, and conformally map the surfaces to canonical domains, or circle domains on canonical spaces. Two surfaces are conformally equivalent if and only if their images on canonical domains are isometric.

Figures 1.4 and 1.5 show the conformal mappings from the surfaces to canonical domains. All genus zero closed surfaces can be conformally mapped to the unit sphere, so are conformally equivalent. All genus one closed surfaces can be mapped to a flat torus, $\mathbb{E}^2/\Lambda$, which is the Euclidean plane $\mathbb{E}^2$ quotient a lattice Λ,

$$\Lambda = \{m + n\omega,\ m, n \in \mathbb{Z}\}.$$

The lattice parameter $\omega \in \mathbb{C}$ is the total conformal module. Two points $p, q \in \mathbb{E}^2$ are equivalent if and only if $p - q \in \Lambda$. All high genus closed surfaces with hyperbolic metric can be represented as $\mathbb{H}^2/\Lambda$, where Λ is a subgroup of hyperbolic isometry group, the so-called *Möbius transformation group*. The group Λ is finitely generated by $2g$ generators. The conformal module is given by these generators. Similarly, compact metric surfaces with boundaries can be deformed to the circle domains in canonical spaces,

$$\{\mathbb{S}^2, \mathbb{E}^2, \mathbb{H}^2\}/\Lambda - \cup_k D_k,$$

where D_k's are circles. The generators of Λ and the centers and radii of D_k's form the conformal module of the surface.

The computational algorithms are based on Ricci flow theory in geometric analysis and Riemann surface theory.

Isometry Group

If two surfaces $(S_1, \mathbf{g}_1), (S_2, \mathbf{g}_2)$ are isometrically equivalent, then they must be conformally equivalent. Let $f_k : S_k \rightarrow D, k = 1, 2$ be the conformal mappings induced by Ricci flow, where D is the canonical space. Let the Riemannian metric on D be $\mathbf{g}_0$, and

$$f_1^* \mathbf{g}_0 = e^{2\lambda_1} \mathbf{g}_1,\ \ f_2^* \mathbf{g}_0 = e^{2\lambda_2} \mathbf{g}_2.$$

Two surfaces are isometric if and only if we can choose f_1 and f_2, such that $\lambda_1 \equiv \lambda_2$.

The computational algorithms are based on surface differential geometry and Riemannian geometry.

Table 1.1 Surface hierarchical geometric structures

Geometric structure	Transformation	Geometry	Main representation
Second fundamental form	Rigid motion	Differential geometry	Mean curvature H
Riemannian metric	Isometry	Riemannian geometry	Conformal factor λ
Conformal structure	Conformal mapping	Conformal geometry	Conformal module
Topological structure	Homeomorphism	Topology	Fundamental group π_1

Rigid Motion Group

Suppose two surfaces embedded in Euclidean space, $(S_1, \mathbf{g}_1)$, $(S_2, \mathbf{g}_2)$, differ by a rigid motion, we can find two conformal mappings $f_k : S_k \to D$ which map the surfaces onto the canonical domain, such that the corresponding conformal factor functions λ_k and mean curvature functions H_k are equal,

$$\lambda_1 \circ f_1^{-1} \equiv \lambda_2 \circ f_2^{-1}, \ H_1 \circ f_1^{-1} \equiv H_2 \circ f_2^{-1}.$$

Here the mean curvature can be interpreted as the area variation of the offset surface, which is defined in (1.5).

The computational algorithms are based on surface differential geometry.

Table 1.1 summarizes the geometric structures for a surface embedded in $\mathbb{E}^3$. Each geometric structure corresponds to a geometry and has a special representation. All the geometric structures form a hierarchy, the higher level structures are based on lower level structures, and represented as functions on the latter.

1.6.2   Shape Comparison

By using Ricci flow, a metric surface $(S, \mathbf{g})$ is conformally deformed to the canonical space $f : S \to \mathbb{D}$. The uniformization gives a special surface parameterization, the so-called *isothermal coordinates* (x, y). Under the isothermal coordinates, the Riemannian metric tensor has the simplest form and can be represented by a scalar function, $\mathbf{g} = e^{2\lambda}(dx^2 + dy^2)$. All the geometric operators have the simplest forms under isothermal coordinates, such as the gradient $\nabla_{\mathbf{g}} = e^{-\lambda}\left(\frac{\partial}{\partial x}, \frac{\partial}{\partial y}\right)^T$, the Laplace–Beltrami operator $\Delta_{\mathbf{g}} = e^{-2\lambda}\left(\frac{\partial^2}{\partial x^2} + \frac{\partial^2}{\partial y^2}\right)$. This improves the efficiency for extracting local geometric features, such as the Gauss curvature,

$$K = -\Delta_{\mathbf{g}}\lambda,$$

the mean curvature,

$$H = \langle \Delta_{\mathbf{g}}\mathbf{r}(x, y), \mathbf{n}(x, y)\rangle, \tag{1.5}$$

where $\mathbf{r}(x, y)$ and $\mathbf{n}(x, y)$ are the position vector and the normal vector of the surface, the principle curvatures,

$$k_1, k_2 = H \pm \sqrt{H^2 - K},$$

the shape index, the geodesics, and so on.

The conformal modules induced by Ricci flow are global shape features. The shortest geodesics in each homotopy class under the uniformization metric form the geodesic spectrum, which gives a global shape descriptor as well. The dynamics of the curvatures during Ricci flow can also be applied as a multi-resolution shape descriptor. Furthermore, Ricci flow preserves intrinsic symmetry of the surface. It can be applied for detecting global symmetry under the uniformization metric.

Various distances between two shapes $(S_1, \mathbf{g}_1)$ and $(S_2, \mathbf{g}_2)$ can be defined. Let $f_k : S_k \to \mathbb{D}$ be the uniformization transformation using Ricci flow, then the distance between the two surfaces is given by

$$d(S_1, S_2) = \int_{\mathbb{D}} (\lambda_1 \circ f_1^{-1} - \lambda_2 \circ f_2^{-1})^2 + (H_1 \circ f_1^{-1} - H_2 \circ f_2^{-1})^2 dx dy.$$

The distance is zero if and only if two surfaces differ by a rigid motion. If the first term is zero, then two surfaces are isometric. In practice, we can find an automorphism of the canonical space $\mathbb{D}$, $\phi : \mathbb{D} \to \mathbb{D}$, such that the distance is the minimizer,

$$d(S_1, S_2) = \min_{\phi} \int_{\mathbb{D}} (\lambda_1 \circ f_1^{-1} - \lambda_2 \circ f_2^{-1} \circ \phi)^2 + (H_1 \circ f_1^{-1} - H_2 \circ f_2^{-1} \circ \phi)^2 dx dy.$$

$$(1.6)$$

1.6.3 Surface Registration

Figure 1.9 explains the framework for surface registration. By using Ricci flow, two surfaces are mapped to the uniformization domains, $f_k : S_k \to \mathbb{D}$. We then compute an automorphisim $\phi : \mathbb{D} \to \mathbb{D}$. The registration between two surfaces is given by the composition

$$f_2^{-1} \circ \phi \circ f_1 : S_1 \to S_2.$$

Initially, the mapping ϕ can be chosen as a harmonic mapping between the canonical domains. If both the source and the target are closed genus zero surfaces, then the canonical domain is the unit sphere; harmonic mapping ϕ must be a Möbius transformation. If the input shapes are genus zero surfaces with a single boundary, then the canonical domain is the unit disk; if the boundary mapping is a homeomorphism, then the interior harmonic mapping is diffeomorphic. If the input shapes are genus one closed surfaces, then the canonical domains are flat tori; the

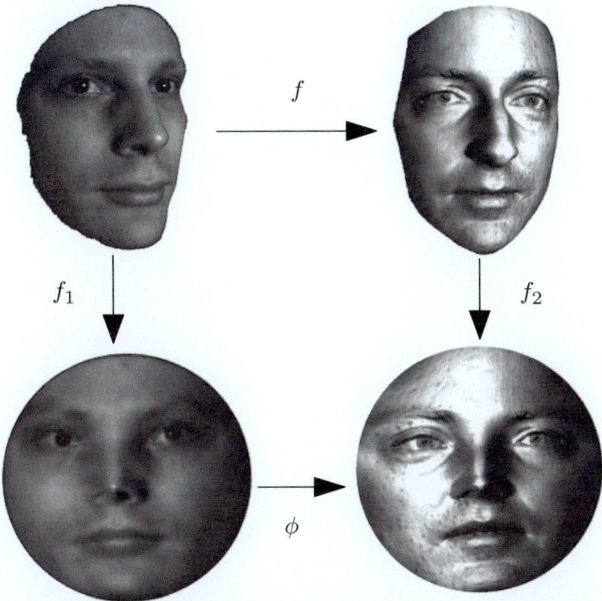

Fig. 1.9 Computational framework for surface registration

harmonic mapping is an affine mapping. If the input surfaces are of high genus, then the canonical domains are hyperbolic surfaces; harmonic mapping between hyperbolic surfaces exists and is unique in each homotopy class and diffeomorphic.

The mapping can be further optimized by minimizing various energies, such as the one defined in (1.6). Other criteria can be added to the energy, such as feature correspondence constraints, smoothness of the mapping, texture consistency, temporal consistency, and prior knowledge about the mappings. The optimization can be performed in the diffeomorphism space of the canonical space. The variational calculus can be performed using quasi-conformal geometric method. If we choose the energy as the distortion of the global conformal structure, then the optimal mapping is the classical Teichmüller map.

Chapter 2
Surface Topology and Geometry

This chapter briefly reviews the fundamental concepts and theorems in algebraic topology [1], surface differential geometry [6], and surface Ricci flow [4, 7]. Detailed discussion on Ricci flow on general Riemannian manifolds can be found in [5]. Advanced topics on differential geometry related to Yamabe equations can be found in [9].

2.1 Surface Topology

Topology studies the invariants under homeomorphism transformation group. Algebraic topology studies the topologies of spaces and the mappings among spaces by algebraic means. Generally, different groups are associated with different spaces, such as fundamental group, homology group, and cohomology group. The structures of these groups convey the topological information about the spaces. The homomorphisms among these groups reflect the properties of the mappings among the spaces. In reality, most surfaces are the boundaries of some finite volumes; therefore, they are compact and orientable. In the following, we focus on the fundamental groups and covering spaces of compact orientable surfaces.

Definition 2.1 (Connected Sum). The connected sum $S_1 \# S_2$ is formed by deleting the interior of disks $D_i \subset S_i$ and attaching the resulting punctured surfaces $S_i - D_i$ to each other by an orientation reversing homeomorphism $h : \partial D_1 \to \partial D_2$, where ∂D_i represents the boundary of D_i. Let $p \in \partial D_1$ and $q \in \partial D_2$. p is equivalent to q, $p \sim q$, if $q = h(p)$. So $S_1 \# S_2 := \{(S_1 - D_1) \cup (S_2 - D_2)\}/ \sim$. See Fig. 2.1.

Theorem 2.1 (Classification for Compact Orientable Surfaces). *Any closed connected orientable surface is homeomorphic to either a sphere or a finite connected sum of tori,*

$$S = \mathbb{S}^2 \# T_1 \# T_2 \cdots \# T_g,$$

W. Zeng and X.D. Gu, *Ricci Flow for Shape Analysis and Surface Registration,*
SpringerBriefs in Mathematics, DOI 10.1007/978-1-4614-8781-4_2,
© Wei Zeng, Xianfeng David Gu 2013

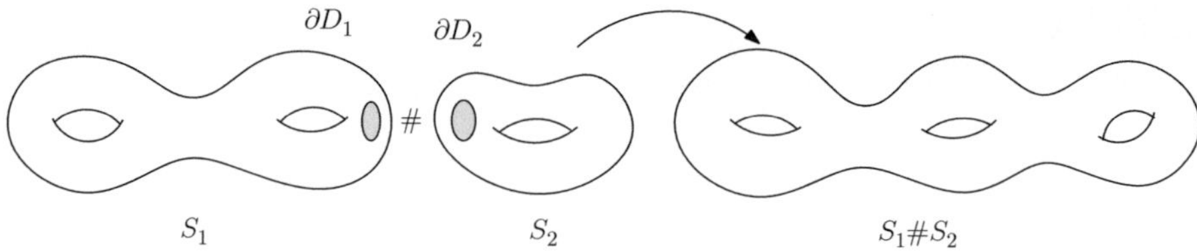

Fig. 2.1 Connected sum

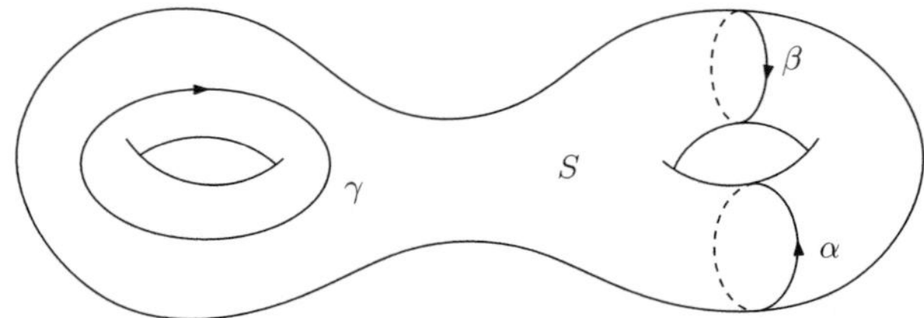

Fig. 2.2 α is homotopic to β, not homotopic to γ

where $\mathbb{S}^2$ is the unit sphere, T_i is a torus, $i = 1, 2, \cdots, g$. g is called the genus of the surface, and each T_i is a handle.

In general, the genus g and the number of boundary components b are the total topological invariants. Surface topology is usually represented by its fundamental group.

2.1.1 Fundamental Group

Definition 2.2 (Homotopy). Two continuous maps $f_0, f_1 : M \to N$ are *homotopic* if there is a continuous map $F : M \times [0, 1] \to N$ such that $F(\cdot, 0) = f_0$ and $F(\cdot, 1) = f_1$. The map F is called a *homotopy* between f_0 and f_1, denoted as $f_0 \cong f_1$ or $F : f_0 \cong f_1$. For each $t \in [0, 1]$, we denote $F(\cdot, t)$ by $f_t : M \to N$, where f_t is a continuous map.

A map $f : [0, 1] \to M$ from the unit interval to a topological space M is called a *path* in M. If f and g are two paths in M with $f(1) = g(0)$, then the *product* of f and g is a path $f \cdot g$, which is defined as

$$f \cdot g(t) = \begin{cases} f(2t) & 0 \le t \le \frac{1}{2} \\ g(2t - 1) & \frac{1}{2} \le t \le 1 \end{cases}. \tag{2.1}$$

Fix a base point $q \in M$, a *loop* with base point q is a path such that $f(0) = f(1) = q$. Two loops on a surface are homotopic to each other, if they can deform to each other without leaving the surface, as shown in Fig. 2.2.

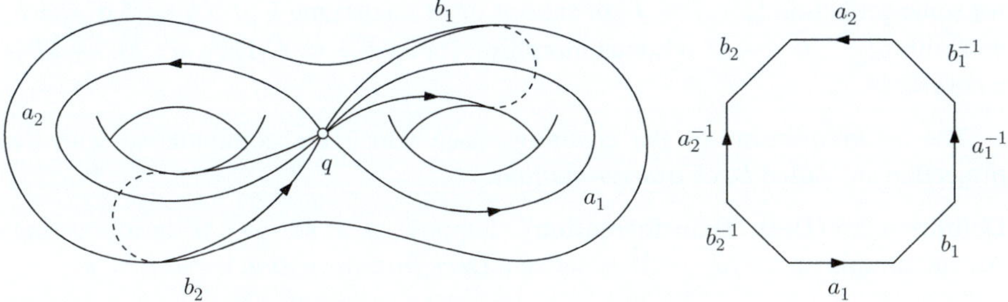

Fig. 2.3 A set of canonical basis of the fundamental group $\pi_1(M, q)$

Definition 2.3 (Fundamental Group). All the homotopy classes of loops with base point q under the product (2.1) form a group, the so-called *fundamental group* of the surface, denoted as $\pi_1(M, q)$.

The fundamental group is finitely generated. Intuitively, each handle T_i is a torus, which is the direct product of two circles, $T_i = \mathbb{S}^1 \times \mathbb{S}^1$. We denote the first circle as a_i, and the second circle b_i, then all such $\{(a_i, b_i)\}$'s are the generators of $\pi_1(M, q)$.

Definition 2.4 (Canonical Fundamental Group Basis). A fundamental group basis $\{a_1, b_1, a_2, b_2, \cdots, a_g, b_g\}$ is canonical, if

(1) a_i and b_i intersect at the same point q.
(2) a_i and a_j, b_i and b_j, a_i and b_j only touch at q, $i \neq j$.

As shown in Fig. 2.3, if we slice the surface along the canonical fundamental group generators, then we will get a $4g$-gon. The boundary is $a_1 b_1 a_1^{-1} b_1^{-1} \cdots a_g b_g a_g^{-1} b_g^{-1}$, which can shrink to a point. For general compact orientable closed surfaces, the following theorem holds.

Theorem 2.2 (Fundamental Groups of General Surfaces). *The fundamental group of the surface* $M = \mathbb{S}^2 \# g \mathbb{T}^2$ *(connected sum of g tori with* $\mathbb{S}^2$*) is the group with generators* $\{a_1, b_1, a_2, b_2, \cdots, a_g, b_g\}$ *and one relation* $\prod_{k=1}^{g} [a_k, b_k] = e$, *where* $[a, b] = aba^{-1}b^{-1}$.

2.1.2 Covering Space

Definition 2.5 (Covering Space). Let $p : \tilde{M} \to M$ be a continuous map and p is onto. Suppose for all $q \in M$, there is an open neighborhood U of q such that

$$p^{-1}(U) = \cup_{j \in J} \tilde{U}_j,$$

for some collection $\{\tilde{U}_j, j \in J\}$ of subsets of $\tilde{M}$, satisfying $\tilde{U}_j \cap \tilde{U}_k = \emptyset$ if $j \neq k$, and with $p|_{\tilde{U}_j} : \tilde{U}_j \to U$ a homeomorphism for each $j \in J$. Then $p : \tilde{M} \to M$ is a *covering*.

The automorphisms of the covering space which are commutative with the projection are called *Deck transformations*.

Definition 2.6 (Deck Transformation). Suppose $p : \tilde{M} \to M$ is a covering. An automorphism $\tau : \tilde{M} \to \tilde{M}$ is called a *Deck transformation* if $p \circ \tau = p$.

All the deck transformations form a group $Deck(\tilde{M})$, called the *deck transformation group*. M is homeomorphic to the quotient space

$$\tilde{M}/Deck(\tilde{M}) \cong M.$$

Definition 2.7 (Fundamental Domain). A closed subset $D \in \tilde{M}$ is called a *fundamental domain* of the $Deck(\tilde{M})$, if $\tilde{M}$ is the union of conjugates of D,

$$\tilde{M} = \bigcup_{\tau \in Deck} \tau D,$$

and the intersection of any two conjugates has no interior.

Among all covering spaces for a given surface, the one with the simplest topology is the so-called *universal covering*.

Definition 2.8 (Universal Covering). Suppose $p : \tilde{M} \to M$ is a covering. If $\tilde{M}$ is simply connected $(\pi(\tilde{M}, \tilde{q}) = \langle e \rangle)$, then the covering is the *universal covering*.

Theorem 2.3 (Universal Covering Space for Surfaces). *The universal covering spaces of orientable closed surfaces are sphere $\mathbb{S}^2$ (genus zero), plane $\mathbb{E}^2$ (genus one), and hyperbolic disk $\mathbb{H}^2$ (high genus).*

Figure 1.4 shows the universal covering spaces of all orientable closed surfaces.

2.2 Surface Differential Geometry

Movable frame and exterior differentiation are the power methods for studying surface differential geometry. We will briefly review the fundamental concepts and theorems for surface differential geometry using the movable frame method, due to its simplicity. A thorough introduction to exterior calculus and movable frame can be found in [3]. We use bold letters to represent vectors, such as $\mathbf{r}$ and $\mathbf{e}$, and Greek letters for differential forms, such as ω and τ.

2.2.1 Movable Frame Method

We apply movable frame method to study surfaces in $\mathbb{E}^3$. Assume the equation for a surface S is $\mathbf{r} = \mathbf{r}(u, v) : D \to \mathbb{E}^3$. Select a smooth orthonormal frame field locally, and at each point $\mathbf{r}(u, v)$ define an orthonormal frame

$$\{\mathbf{r}(u, v); \mathbf{e}_1(u, v), \mathbf{e}_2(u, v), \mathbf{e}_3(u, v)\},$$

such that $\mathbf{e}_3$ is the normal field, $\mathbf{e}_3(u, v) = \mathbf{n}(u, v)$. Taking the exterior derivative of the movable frame $\{\mathbf{r}; \mathbf{e}_1, \mathbf{e}_2, \mathbf{e}_3\}$, we get the surface structure equation

$$d\mathbf{r} = \omega_1 \mathbf{e}_1 + \omega_2 \mathbf{e}_2,$$

$$d \begin{pmatrix} \mathbf{e}_1 \\ \mathbf{e}_2 \\ \mathbf{e}_3 \end{pmatrix} = \begin{pmatrix} 0 & \omega_{12} & \omega_{13} \\ -\omega_{12} & 0 & \omega_{23} \\ -\omega_{13} & -\omega_{23} & 0 \end{pmatrix} \begin{pmatrix} \mathbf{e}_1 \\ \mathbf{e}_2 \\ \mathbf{e}_3 \end{pmatrix}.$$

From $d^2\mathbf{r} = 0$, we obtain

$$d\omega_1 = \omega_{12} \wedge \omega_2, \tag{2.2}$$

$$d\omega_2 = \omega_{21} \wedge \omega_1, \tag{2.3}$$

and

$$\omega_1 \wedge \omega_{13} + \omega_2 \wedge \omega_{23} = 0.$$

From (2.2) and (2.3), we directly obtain

$$\omega_{12} = \frac{d\omega_1}{\omega_1 \wedge \omega_2} \omega_1 + \frac{d\omega_2}{\omega_1 \wedge \omega_2} \omega_2. \tag{2.4}$$

From $d^2\mathbf{e}_1 = 0$, we get the *Gauss equation*

$$d\omega_{12} = \omega_{13} \wedge \omega_{32} \tag{2.5}$$

and *Codazzi equation*

$$d\omega_{13} = \omega_{12} \wedge \omega_{23}. \tag{2.6}$$

Similarly, from $d^2\mathbf{e}_2 = 0$, we obtain another Codazzi equation

$$d\omega_{23} = \omega_{21} \wedge \omega_{13}. \tag{2.7}$$

2.2.2 First and Second Fundamental Forms

The first fundamental form of the surface is given by

$$I = \langle d\mathbf{r}, d\mathbf{r} \rangle = \langle \omega_1 \mathbf{e}_1 + \omega_2 \mathbf{e}_2, \omega_1 \mathbf{e}_1 + \omega_2 \mathbf{e}_2 \rangle = \omega_1^2 + \omega_2^2. \qquad (2.8)$$

The second fundamental form is given by

$$II = -\langle d\mathbf{r}, d\mathbf{e}_3 \rangle = -\langle \omega_1 \mathbf{e}_1 + \omega_2 \mathbf{e}_2, \omega_{31} \mathbf{e}_1 + \omega_{32} \mathbf{e}_2 \rangle = \omega_1 \omega_{13} + \omega_2 \omega_{23}. \quad (2.9)$$

The first and second fundamental forms are not independent, but related by the Gauss and Codazzi equations. Further, Gauss–Codazzi equations are also sufficient conditions for the existence and the uniqueness of a surface.

Let Ω be a matrix. Each entry is a differential form,

$$\Omega = \begin{pmatrix} 0 & \omega_{12} & \omega_{13} \\ -\omega_{12} & 0 & \omega_{23} \\ -\omega_{13} & -\omega_{23} & 0 \end{pmatrix}.$$

Equations (2.2) and (2.3) can be summarized as

$$d \begin{pmatrix} \omega_1 & \omega_2 & 0 \end{pmatrix} = \begin{pmatrix} \omega_1 & \omega_2 & 0 \end{pmatrix} \wedge \Omega,$$

where the wedge product can be interpreted as the matrix product, and the product of two entries is replaced by the wedge product of two differential forms. Similarly, the Gauss–Codazzi equations (2.5), (2.6) and (2.7) can be summarized as

$$d\Omega = \Omega \wedge \Omega.$$

The fundamental theorem for surface differential geometry is as follows.

Theorem 2.4. *Suppose $D \subset \mathbb{R}^2$ is a domain on the (u, v) plane. Given 5 differential 1-forms, $\omega_1, \omega_2, \omega_{12}, \omega_{13}, \omega_{23}$, satisfying*

$$d \begin{pmatrix} \omega_1 & \omega_2 & 0 \end{pmatrix} = \begin{pmatrix} \omega_1 & \omega_2 & 0 \end{pmatrix} \wedge \Omega,$$
$$d\Omega = \Omega \wedge \Omega,$$

then for any given initial condition at $(u_0, v_0) \in D$, the position $\mathbf{r}(u_0, v_0)$ and an orthonormal frame $\{\mathbf{r}; \mathbf{e}_1, \mathbf{e}_2, \mathbf{e}_3\}(u_0, v_0)$, there is a unique surface patch $\mathbf{r}(u, v)$ with an orthonormal frame field $\{\mathbf{r}; \mathbf{e}_1, \mathbf{e}_2, \mathbf{e}_3\}(u, v)$ in a neighborhood of (u_0, v_0), such that

$$d\mathbf{r} = \begin{pmatrix} \omega_1 & \omega_2 & 0 \end{pmatrix} \begin{pmatrix} \mathbf{e}_1 \\ \mathbf{e}_2 \\ \mathbf{e}_3 \end{pmatrix}$$

and

$$
d \begin{pmatrix} \mathbf{e}_1 \\ \mathbf{e}_2 \\ \mathbf{e}_3 \end{pmatrix} = \Omega \begin{pmatrix} \mathbf{e}_1 \\ \mathbf{e}_2 \\ \mathbf{e}_3 \end{pmatrix}.
$$

The proof can be found in classical differential geometry textbook, such as [6].

Because $d\mathbf{r} = \omega_1\mathbf{e}_1 + \omega_2\mathbf{e}_2$, the area element of the surface is $\omega_1 \wedge \omega_2$. Similarly, because $d\mathbf{e}_3 = \omega_{31}\mathbf{e}_1 + \omega_{32}\mathbf{e}_2$, the area element of the unit sphere is $\omega_{31} \wedge \omega_{32}$. Then the determinant of the Jacobian matrix of the Weingarten map $d\mathbf{r} \to d\mathbf{e}_3$ is given by

$$
K = \frac{\omega_{31} \wedge \omega_{32}}{\omega_1 \wedge \omega_2}.
$$

From Gauss equation (2.5), we get the important equation for Gauss curvature,

$$
d\omega_{12} = -\omega_{31} \wedge \omega_{32} = -K\omega_1 \wedge \omega_2. \tag{2.10}
$$

We say a geometric quantity is intrinsic, if it is solely determined by the first fundamental form, namely, Riemannian metric. From (2.4) and (2.10), we see that Gauss curvature K is solely determined by ω_1 and ω_2; therefore, we obtain Gauss' *Theorema Egregium*.

Theorem 2.5. *The Gaussian curvature solely depends on the first fundamental form, namely, is intrinsic.*

Therefore, the Gaussian curvature of a surface is invariant under local isometry.

2.2.3 Curves on Surfaces

Consider a curve C on a surface S with a local representation $C : (u(s), v(s))$, where s is the arc length parameter. Let $\boldsymbol{\alpha}$ be the tangent direction of C, and $\theta(s)$ be the angle from $\mathbf{e}_1$ to $\boldsymbol{\alpha}$. By direct computation, the *geodesic curvature* is given by

$$
k_g = \frac{d\theta + \omega_{12}}{ds}. \tag{2.11}
$$

If $k_g \equiv 0$, then the curve is called a *geodesic*. The *normal curvature* is

$$
k_n = \frac{\omega_1\omega_{13} + \omega_2\omega_{23}}{ds^2} = \frac{II}{I}.
$$

At each point, there are two orthogonal tangent directions, along which the normal curvature reaches the minimum k_1 and maximum k_2. k_1 and k_2 are called the *principle curvatures*, and the two directions are called the *principle directions*, as shown in Fig. 2.4. The Gauss curvature is the product of principle curvatures, $K = k_1 k_2$; the *mean curvature* is the mean value of the principle curvatures, $H = (k_1 + k_2)/2$.

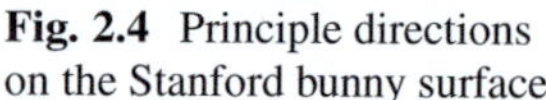

Fig. 2.4 Principle directions
on the Stanford bunny surface

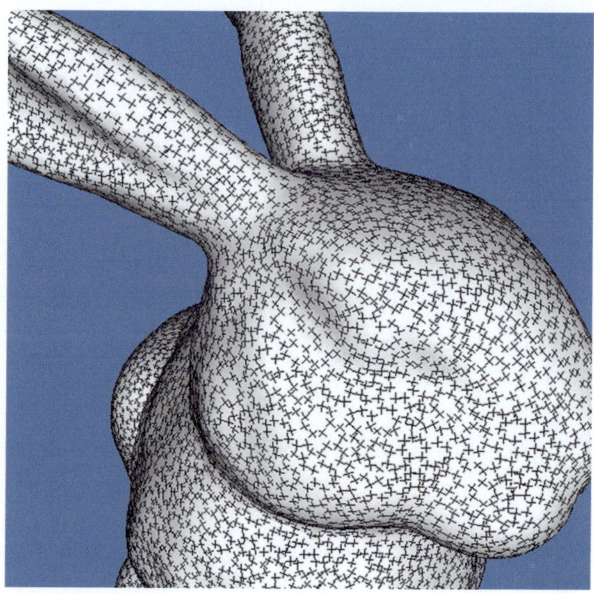

From (2.10) and (2.11), we can prove the Gauss–Bonnet theorem, which claims
that although the Gauss curvature is determined by the Riemannian metric, the total
curvature is solely determined by the surface topology.

Theorem 2.6 (Gauss–Bonnet). *Suppose S is a compact two-dimensional Riemannian manifold with piecewise-smooth boundary ∂S. Let K be the Gauss curvature,
k_g the geodesic curvature of ∂S, and $\theta_k, k = 1, 2 \cdots, n$ be the exterior angles of
∂S. Then*

$$\int_S K dA + \int_{\partial S} k_g ds + \sum_{k=1}^{n} \theta_k = 2\pi \chi(S),$$

where $\chi(S)$ is the Euler characteristics of the surface.

2.3 Conformal Metric Deformation

2.3.1 Isothermal Coordinates

Given a metric surface, one can choose *isothermal coordinates* to facilitate geometric computations, as shown in Fig. 2.5. Most differential operators, such as
gradient and Laplace–Beltrami operators, have the simplest form under isothermal
coordinates.

Fig. 2.5 Isothermal coordinate system on the Stanford bunny surface. The mapping from the surface to the parameter plane is conformal, which preserves angles and infinitesimal circles

Definition 2.9 (Isothermal Coordinates). On a surface S with a Riemannian metric $\mathbf{g}$, a local coordinates system (u, v) is an isothermal coordinate system, if

$$\mathbf{g}(u, v) = e^{2\lambda(u,v)}(du^2 + dv^2), \tag{2.12}$$

where $\lambda : S \to \mathbb{R}$ is a function defined on the surface, and called *conformal factor*.

Isothermal coordinates on metric surfaces always exist, which can be proved either using surface Ricci flow or quasi-conformal mapping. In the later part, we give a proof by solving a Beltrami equation. An elementary proof can be found in Chern's work [2].

Theorem 2.7 (Existence of Isothermal Coordinates). *Let $(S, \mathbf{g})$ be a compact orientable surface, then every point of S has a neighborhood whose local coordinates are isothermal parameters.*

2.3.2   Gauss Curvature Under Conformal Deformation

We use movable frame method to deduce Gauss curvature under isothermal coordinates. Let $(S, \mathbf{g})$ be a surface embedded in $\mathbb{E}^3$, with position vector function $\mathbf{r}(u, v)$ and isothermal coordinates (u, v). We denote $\partial_u \mathbf{r}$ as $\mathbf{r}_u$, and $\partial_v \mathbf{r}$ as $\mathbf{r}_v$, then

$$\langle \mathbf{r}_u, \mathbf{r}_u \rangle = e^{2\lambda}, \ \langle \mathbf{r}_v, \mathbf{r}_v \rangle = e^{2\lambda}, \ \langle \mathbf{r}_u, \mathbf{r}_v \rangle = 0.$$

Choose an orthonormal frame

$$\mathbf{e}_1 = e^{-\lambda}\mathbf{r}_u, \ \mathbf{e}_2 = e^{-\lambda}\mathbf{r}_v, \ \mathbf{e}_3 = \mathbf{n}.$$

Then

$$d\mathbf{r} = \mathbf{r}_u du + \mathbf{r}_v dv = \mathbf{e}_1 e^{\lambda} du + \mathbf{e}_2 e^{\lambda} dv = \omega_1 \mathbf{e}_1 + \omega_2 \mathbf{e}_2,$$

where $\omega_1 = e^{\lambda} du$ and $\omega_2 = e^{\lambda} dv$. From (2.4), we get $\omega_{12} = -\lambda_v du + \lambda_u dv$. Therefore

$$d\omega_{12} = (\lambda_{vv} + \lambda_{uu}) du \wedge dv = -K\omega_1 \wedge \omega_2 = -Ke^{2\lambda} du \wedge dv,$$

then we obtain

$$K(u, v) = -e^{-2\lambda(u,v)} \left(\frac{\partial^2}{\partial u^2} + \frac{\partial^2}{\partial v^2} \right) \lambda = -\Delta_{\mathbf{g}}\lambda, \qquad (2.13)$$

where the Laplace–Beltrami operator is

$$\Delta_{\mathbf{g}} = e^{-2\lambda(u,v)} \left(\frac{\partial^2}{\partial u^2} + \frac{\partial^2}{\partial v^2} \right).$$

Let $\bar{\mathbf{g}}$ be another Riemannian metric, conformal to the original metric,

$$\bar{\mathbf{g}} = e^{2\tau}\mathbf{g},$$

where τ is a scalar function. We choose isothermal coordinates for both $\mathbf{g}$ and $\bar{\mathbf{g}}$, then

$$\mathbf{g} = e^{2\lambda}(du^2 + dv^2), \ \bar{\mathbf{g}} = e^{2(\lambda+\tau)}(du^2 + dv^2),$$

The Gauss curvature $\bar{K}$ induced by $\bar{\mathbf{g}}$ becomes

$$\bar{K} = -e^{-2(\lambda+\tau)}\Delta(\lambda + \tau) = e^{-2\tau}(-e^{-2\lambda}\Delta\lambda - e^{-2\lambda}\Delta\tau) = e^{-2\tau}(K - \Delta_{\mathbf{g}}\tau).$$

So we obtain the Yamabe equation

$$\bar{K} = e^{-2\tau}(K - \Delta_{\mathbf{g}}\tau). \qquad (2.14)$$

2.3.3 *Geodesic Curvature Under Conformal Deformation*

Suppose C is a curve on the surface and the tangent direction $\boldsymbol{\alpha}$ of the curve has the angle θ to the $\mathbf{e}_1$ direction. Then the geodesic curvature of C is

$$k_g = \frac{d\theta + \omega_{12}}{ds}.$$

Choose the isothermal coordinates (u, v),

$$\omega_{12} = \lambda_v du - \lambda_u dv,$$

$$ds = e^\lambda \sqrt{du^2 + dv^2},$$

$$\frac{du}{ds} = e^{-\lambda} \cos\theta, \quad \frac{dv}{ds} = e^{-\lambda} \sin\theta,$$

$$k_g = \frac{d\theta}{ds} + \frac{\lambda_v du - \lambda_u dv}{ds} = \frac{d\theta}{ds} - e^{-\lambda}(\lambda_u \sin\theta - \lambda_v \cos\theta)$$

$$= \frac{d\theta}{ds} - \langle \nabla_{\mathbf{g}} \lambda, \mathbf{n} \rangle,$$

where $\nabla_{\mathbf{g}} = e^{-\lambda}\left(\frac{\partial}{\partial u}, \frac{\partial}{\partial v}\right)$, and $\mathbf{n} = (\sin\theta, -\cos\theta)$ is the outward normal of the curve on the tangent plane. The geodesic curvature can also be written as

$$k_g = \frac{d\theta}{ds} - \partial_{\mathbf{n},\mathbf{g}} \lambda.$$

Assume $\bar{\mathbf{g}}$ is another Riemannian metric conformal to $\mathbf{g}$, $\bar{\mathbf{g}} = e^{2\tau}\mathbf{g}$. Choose the isothermal coordinates (u, v),

$$\bar{\mathbf{g}} = e^{2(\tau+\lambda)}(du^2 + dv^2), \mathbf{g} = e^{2\lambda}(du^2 + dv^2),$$

then

$$\bar{k}_g = \frac{d\theta}{d\bar{s}} - \partial_{\mathbf{n},\bar{\mathbf{g}}}(\lambda + \tau).$$

Because $d\bar{s} = e^\tau ds$,

$$\frac{d\theta}{d\bar{s}} = e^{-\tau}\frac{d\theta}{ds}.$$

Because $\nabla_{\bar{\mathbf{g}}} = e^{-\tau-\lambda}\left(\frac{\partial}{\partial u}, \frac{\partial}{\partial v}\right) = e^{-\tau}\nabla_{\mathbf{g}}$,

$$\partial_{\mathbf{n},\bar{\mathbf{g}}} = \langle \nabla_{\bar{\mathbf{g}}}, \mathbf{n} \rangle = e^{-\tau}\langle \nabla_{\mathbf{g}}, \mathbf{n} \rangle = e^{-\tau}\partial_{\mathbf{n},\mathbf{g}}.$$

Therefore

$$\bar{k}_g = e^{-\tau}\frac{d\theta}{ds} - e^{-\tau}\partial_{\mathbf{n},\mathbf{g}}(\lambda + \tau) = e^{-\tau}\left(\frac{d\theta}{ds} - \partial_{\mathbf{n},\mathbf{g}}\lambda - \partial_{\mathbf{n},\mathbf{g}}\tau\right) = e^{-\tau}(k_g - \partial_{\mathbf{n},\mathbf{g}}\tau).$$

Theorem 2.8 (Surface Yamabe Problem). *Suppose S is a surface with a Riemannian metric $\mathbf{g}$, which induces Gauss curvature K and geodesic curvature k_g on the boundary. Let*

$$\bar{\mathbf{g}} = e^{2\lambda}\mathbf{g}$$

*be another metric conformal to the original one, which induces Gauss curvature
$\bar{K}$ and geodesic curvature $\bar{k}_g$. Then the changes of Gauss curvature and geodesic
curvature associated with the conformal metric change respectively are*

$$\bar{K} = e^{-2\lambda}(K - \Delta_{\mathbf{g}}\lambda),$$
$$\bar{k}_g = e^{-\lambda}(k_g - \partial_{\mathbf{n},\mathbf{g}}\lambda).$$

Surface Yamabe problem is to find the conformal factor λ from the prescribed
curvatures $\bar{K}$ and $\bar{k}_g$, which can be solved using surface Ricci flow.

2.4   Surface Ricci Flow

Given an n dimensional Riemannian manifold M with metric tensor $\mathbf{g} = (g_{ij})$, the
normalized Ricci flow is defined by the geometric evolution equation

$$\partial_t \mathbf{g}(t) = -2Ric(\mathbf{g}(t)) + \rho\mathbf{g}(t).$$

where Ric is the Ricci curvature tensor and ρ is the mean value of the scalar
curvature,

$$\rho = \frac{2}{n}\frac{\int_M R_{\mathbf{g}}d\mu_{\mathbf{g}}}{\int_M d\mu_{\mathbf{g}}},$$

where $R_{\mathbf{g}}$ and $\mu_{\mathbf{g}}$ are the scalar curvature and the volume element with respect to
the evolving metric $\mathbf{g}(t)$, respectively. Recall that a one-parameter family of metrics
$\{\mathbf{g}(t)\}$, where $t \in [0, T)$ for some $0 < T \le \infty$, is called a solution to the normalized
Ricci flow if it satisfies the above equation at all $p \in M$ and $t \in [0, T)$.

In two dimensions, the Ricci curvature for a metric $\mathbf{g}$ is equal to $\frac{1}{2}R\mathbf{g}$, where R is
the scalar curvature (or twice the Gauss curvature). Therefore, the normalized Ricci
flow equation for surfaces takes the form

$$\partial_t \mathbf{g}(t) = (\rho - R(t))\mathbf{g}(t), \tag{2.15}$$

where ρ is the mean value of the scalar curvature,

$$\rho = \frac{4\pi\chi(M)}{A(0)},$$

where $\chi(M)$ is the Euler characteristic number of M, and $A(0)$ is the total area of
the surface M at time $t = 0$.

Let $(g^{ij}) = (g_{ij})^{-1}$ be the inverse of the matrix (g_{ij}). Set the area element with
respect to metric $\mathbf{g}$ to be

$$\mu_{\mathbf{g}} = \sqrt{det g_{ij}}.$$

Then along the Ricci flow, we compute

$$\partial_t \mu_{\mathbf{g}} = \frac{1}{2} g^{ij} \partial_t g_{ij} \mu_{\mathbf{g}} = (\rho - R)\mu_{\mathbf{g}}.$$

For the total area $A = \int_M d\mu$, we have

$$\partial_t A(t) = \int_M (\rho - R) d\mu_{\mathbf{g}} = 0.$$

Therefore, the normalized Ricci flow preserves the total area, $A(t) = A(0), \forall t > 0$. During the Ricci flow (2.15), the metric deforms conformally, $\mathbf{g}(t) = e^{2\lambda(t)}\mathbf{g}(0)$,

$$\partial_t \lambda = \frac{1}{2}(\rho - R), \ \lambda(0) = 0, \tag{2.16}$$

and from Yamabe equation (2.14),

$$\Delta_0 \lambda - \frac{1}{2}R_0 + \frac{1}{2}Re^{2\lambda} = \Delta_0 \lambda - K_0 + Ke^{2\lambda} = 0.$$

We obtain the curvature evolution equation

$$\partial_t R = \Delta_{g(t)} R + R(R - \rho). \tag{2.17}$$

Let $u = e^{2\lambda}$, then we get the evolution equation for u,

$$\partial_t u = (\rho - R)u.$$

Plug in $R = u^{-1}(R_0 - \Delta_0 \log u)$, we get an evolution flow for u,

$$\partial_t u = \Delta_0 \log u + \rho u - R_0, \ u(0) = u_0. \tag{2.18}$$

For most evolution equations, one proved that the solutions exist for all $t \geq 0$ by combining a short-time existence (and uniqueness) result with prior bounds which show that the solutions cannot develop singularities in finite time. Equation (2.18) is a parabolic equation. It can be set up as a fixed point problem for a contraction mapping. The mapping is obtained by applying the fundamental solution of the linearization at any given $u_0 > 0$ to (2.18); it is a contraction on any sufficiently short-time interval. This gives the proof for the short-time existence of the solution. The long-time existence can be obtained by estimating both the lower and upper bounds of $R(t)$ and $u(t)$, which requires a generalization of Li-Yau's Harnack inequality [8]. The proofs require advanced background knowledge and sophisticated geometric skills, which is beyond the scope of the current book. Details can be found in Hamilton's [7] and Chow's [4] works.

Theorem 2.9 (Hamilton [7]). *Let (M^2, g_0) be compact. If $\rho \leq 0$, or if $R(0) \geq 0$ on all of M^2, then the solution to (2.15) exists for all $t \geq 0$ and converges to a metric of constant curvature.*

Theorem 2.10 (Chow [4]). *If g_0 is any metric on $\mathbb{S}^2$, then its evolution under (2.15) develops positive scalar curvature in finite time, and hence by Theorem 2.9 converges to the round metric as t goes to ∞.*

In Chap. 4, we will give a discrete version of surface Ricci flow theory and prove its convergence (Theorem 4.6). Discrete surface Ricci flow shares the same theoretical framework, the same fundamental principles, and even the same formulae (comparing (2.16) and (4.10)), but only requires elementary geometric knowledge to prove. Furthermore, discrete surface Ricci flow theory leads to the practical computational algorithms directly.

References

1. Armstrong, M.: Basic Topology. Undergraduate Texts in Mathematics. Springer, New York (1983)
2. Chern, S.: An elementary proof of the existence of isothermal parameters on a surface. Proc. Am. Math. Soc. **6**(5), 771–782 (1955)
3. Chern, S.S., Cesari, L.: Global Differential Geometry. Mathematical Association of America, Providence, RI (1989)
4. Chow, B.: The Ricci flow on the 2-sphere. J. Differ. Geom. **33**(2), 325–334 (1991)
5. Chow, B., Lu, P., Ni, L.: Hamilton's Ricci Flow, Graduate Studies in Mathematics, vol. 77. American Mathematical Society, Providence, RI (2006)
6. DoCarmo, M.P.: Differential Geometry of Curves and Surfaces, 1st edn. Pearson (1976)
7. Hamilton, R.: Ricci flow on surfaces. Mathematics and General Relativity. Contemporary Mathematics, vol. 71, pp. 237–261. AMS, Providence, RI (1988)
8. Li, P., Yau, S.T.: On the parabolic kernel of the Schrödinger operator. Acta Math. **156**, 153–201 (1986)
9. Schoen, R., Yau, S.T.: Lecture on Differential Geometry, vol. 1. International Press Incorporated, Boston (1994)

Chapter 3
Riemann Surface

Riemann surface theory studies the invariants under conformal transformation group. This chapter briefly introduces the Riemann surface theory [7], including quasi-conformal mapping [1], Teichmüller space [3,12], and surface harmonic maps [10,11]. Finally, the Teichmüller theory of harmonic maps [13] is covered.

Surfaces are classified by conformal transformation group, and the space of all the orbits is the Teichmüller space, which is a natural model for the shape space. General diffeomorphisms are quasi-conformal maps and can be represented by their Beltrami coefficients. The functional space of all Beltrami coefficients is a natural model for the mapping space. In order to obtain diffeomorphisms, harmonic maps are often applied as the initial mappings, due to their simplicity and stability. Furthermore, under mild conditions, harmonic maps are diffeomorphic and unique in their homotopy classes.

3.1 Conformal Structure

Definition 3.1 (Holomorphic Function). Suppose $f : \mathbb{C} \to \mathbb{C}$ is a complex function, $w = f(z)$, in real form, $f : x + iy \to u(x, y) + iv(x, y)$. If f satisfies the Riemann–Cauchy equation

$$\frac{\partial u}{\partial x} = \frac{\partial v}{\partial y}, \ \frac{\partial u}{\partial y} = -\frac{\partial v}{\partial x},$$

then f is a holomorphic function.

For convenience, we use complex differential operators,

$$dz = dx + i\,dy, \ d\bar{z} = dx - i\,dy,$$

W. Zeng and X.D. Gu, *Ricci Flow for Shape Analysis and Surface Registration*, SpringerBriefs in Mathematics, DOI 10.1007/978-1-4614-8781-4_3, © Wei Zeng, Xianfeng David Gu 2013

Fig. 3.1 Biholomorphic functions are conformal mappings

$$\frac{\partial}{\partial z} = \frac{1}{2}\left(\frac{\partial}{\partial x} - i\frac{\partial}{\partial y}\right), \quad \frac{\partial}{\partial \bar{z}} = \frac{1}{2}\left(\frac{\partial}{\partial x} + i\frac{\partial}{\partial y}\right).$$

If $\frac{\partial f}{\partial \bar{z}} = 0$, then f is holomorphic.

Definition 3.2 (Biholomorphic Function). Suppose $f : \mathbb{C} \to \mathbb{C}$ is invertible and both f and f^{-1} are holomorphic. Then f is a biholomorphic function.

Let $w = f(z)$ be a biholomorphic function. The pullback metric induced by $f : (\mathbb{C}, |dz|^2) \to (\mathbb{C}, |dw|^2)$ is

$$f^*|dw|^2 = |dw|^2 = |w_z dz + w_{\bar{z}} d\bar{z}|^2 = |w_z|^2|dz|^2.$$

Therefore, biholomorphic mappings are conformal diffeomorphisms. Figure 3.1 shows a biholomorphic function. In the figure, we can see that the mapping preserves local shapes, such as the decahedra and the paper bunny models.

Definition 3.3 (Conformal Atlas). Suppose S is a topological surface (two-dimensional manifold) and $\mathscr{A}$ is an atlas, such that all the chart transition functions $\phi_{\alpha\beta} : \mathbb{C} \to \mathbb{C}$ are biholomorphic. Then $\mathscr{A}$ is called a conformal atlas.

Suppose S is a topological surface and $\mathscr{A}_1, \mathscr{A}_2$ are two conformal atlases. If their union $\mathscr{A}_1 \cup \mathscr{A}_2$ is still a conformal atlas, then we say $\mathscr{A}_1$ and $\mathscr{A}_2$ are compatible, denoted as $\mathscr{A}_1 \sim \mathscr{A}_2$. The compatible relation among conformal atlases is an equivalence relation, which can classify all conformal atlases.

Definition 3.4 (Conformal Structure). Suppose S is a topological surface. Consider that all the conformal atlases on M are classified by the compatible relation,

$$\{all\ conformal\ atlases\}/ \sim .$$

Then each equivalence class is called a *conformal structure*.

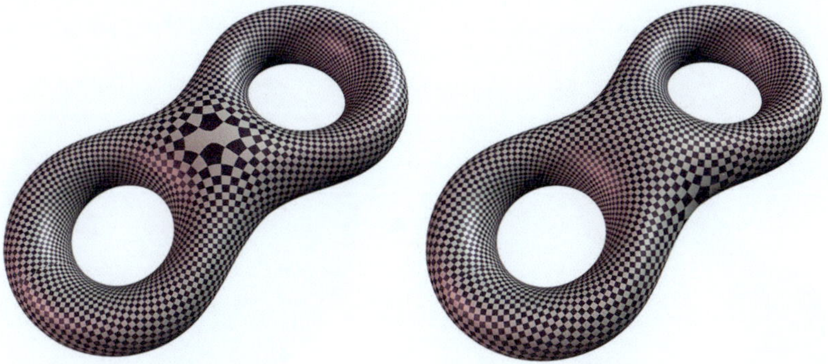

Fig. 3.2 Holomorphic 1-form basis for a genus two surface

Definition 3.5 (Riemann Surface). Suppose S is a surface with a conformal structure. Then S is a Riemann surface.

Theorem 3.1. *All compact orientable metric surfaces are Riemann surfaces.*

Proof. For each point, there exists a neighborhood, where isothermal coordinates exist. The collection of isothermal coordinates charts is a conformal atlas. □

Definition 3.6 (Holomorphic Differential 1-form). A holomorphic differential 1-form on a Riemann surface is a differential form ω, with local presentation $\omega = \phi(z)dz$, where $\phi(z)$ is a holomorphic function. ωdz is invariant under the parameter change.

All the holomorphic 1-forms on a genus g Riemann surface S form a vector space $\Omega(S)$, which is isomorphic to the first cohomology group $H^1(S, \mathbb{R})$, and g complex dimension. The basis for $\Omega(S)$ can be computed using Hodge theory. Details can be found in [4]. Figure 3.2 shows the basis of holomorphic differentials on a genus two Riemann surface.

Definition 3.7 (Holomorphic Quadratic Differential). A holomorphic quadratic differential on a Riemann surface has local presentation $\omega = \phi(z)dz^2$, where $\phi(z)$ is a holomorphic function. ω is invariant under the parameter change.

The complex linear space of all holomorphic quadratic differentials is denoted as $QD(S)$, which is $3g - 3$ complex dimensional

3.2 Teichmüller Space

Definition 3.8 (Conformal Mapping). Suppose S_1 and S_2 are two Riemann surfaces, with conformal structures $\{(U_\alpha, \phi_\alpha)\}$ and $\{(V_\beta, \tau_\beta)\}$, respectively. f is a mapping from S_1 to S_2, $f : S_1 \rightarrow S_2$. If all the local representations $\tau_\beta \circ f \circ \phi_\alpha^{-1}$ are biholomorphic, then we say f is a *biholomorphic mapping* or *conformal mapping* between S_1 and S_2, and S_1 and S_2 are *conformally equivalent*.

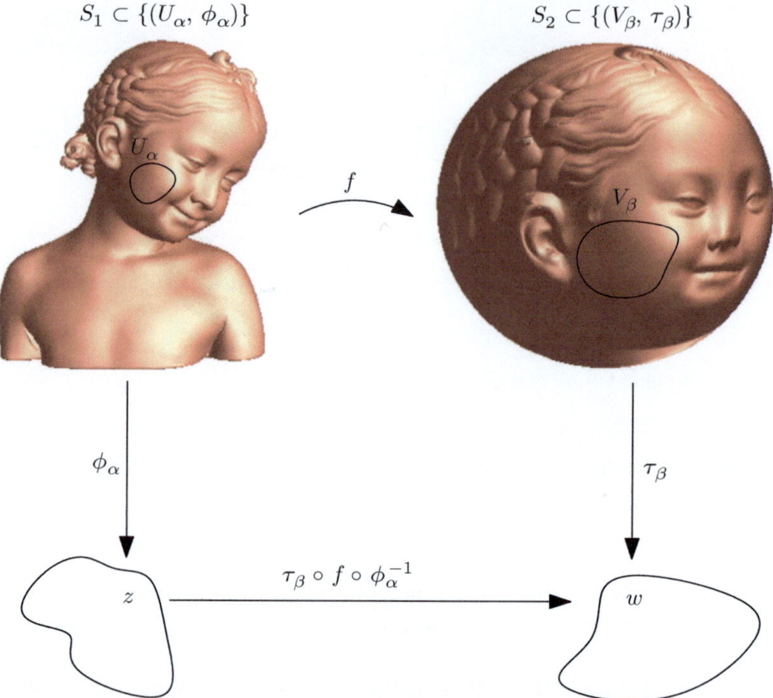

Fig. 3.3 Biholomorphic mapping (conformal mapping) between two Riemann surfaces. All $\tau_\beta \circ f \circ \phi_\alpha^{-1}$'s are biholomorphic

Figure 3.3 illustrates a biholomorphic mapping between two Riemann surfaces.

Suppose S is a compact orientable surface with genus g, and μ_g is the set of all possible conformal structures on S. We say two elements $\mu_1, \mu_2 \in \mu_g$ are equivalent, denoted as $\mu_1 \sim \mu_2$, if there exists a conformal mapping $f : S_{\mu_1} \to S_{\mu_2}$, where S_{μ_1} and S_{μ_2} are Riemann surfaces formed by S equipped with the conformal structures μ_1 and μ_2. The equivalence class of $\mu \in \mu_g$ is denoted as $[\mu]$.

Definition 3.9 (Moduli Space). The set of all equivalence classes $[\mu]$ is called the moduli space of genus g Riemann surface, denoted as R_g,

$$R_g = \mu_g / \sim .$$

The topology of moduli space is complicated. In practice, Teichmüller space is more preferred, which is the universal covering space of the moduli space.

Two conformal structures $\mu_1, \mu_2 \in \mu_g$ are *Teichmüller equivalent*, denoted as $\mu_1 \approx \mu_2$, if there is a conformal mapping $f : S_{\mu_1} \to S_{\mu_2}$. Furthermore, treated as an automorphism of S, $f : S \to S$ is homotopic to the identity.

Definition 3.10 (Teichmüller Space). The set of Teichmüller equivalence classes in μ_g is called the Teichmüller space, denoted as T^g,

$$T^g = \mu_g/\approx .$$

3.3 Conformal Module

In this section, we apply surface Ricci flow to explain the concept of conformal module and analyze the dimension of the corresponding Teichmüller space. We use $T^{(g,b)}$ to represent the Teichmüller space of surfaces of genus g with b boundary components. The constructive proofs can be converted to computational methods directly.

3.3.1 Topological Sphere

A topological sphere (a genus zero closed surface) can be conformally mapped to the unit sphere. The mapping is not unique. Two such conformal mappings differ by a spherical Möbius transformation. Let $\phi : \mathbb{S}^2 \to \bar{\mathbb{C}}$ be the stereographic projection, which maps the unit sphere $\mathbb{S}^2$ to the extended complex plane $\bar{\mathbb{C}} = \mathbb{C} \cup \{\infty\}$,

$$\phi(x, y, z) = \left(\frac{x}{1-z}, \frac{y}{1-z} \right).$$

Let τ be the Möbius transformation from the extended plane to itself,

$$\tau(z) = \frac{az + b}{cz + d}, \ ad - bc = 1, \ a, b, c, d \in \mathbb{C}.$$

Then a spherical Möbius transformation has the form $\phi^{-1} \circ \tau \circ \phi$, which is 6 dimensional. Therefore, all topological spheres are conformally equivalent; the Teichmüller space of all genus zero closed surfaces consists of a single point, $T^{(0,0)} = \{p\}$. Figure 3.4 shows an example, which can be computed using Ricci flow or the spherical harmonic map method. According to Theorem 3.9, harmonic maps between genus zero closed surfaces are conformal.

3.3.2 Topological Quadrilateral

Suppose S is a genus zero surface with a single boundary and four marked boundary points, p_1, p_2, p_3, p_4, sorted counter-clockwisely. Then S is called a *topological quadrilateral* and denoted as $Q(p_1, p_2, p_3, p_4)$. There exists a unique conformal map $\phi : S \to \mathbb{C}$, such that ϕ maps Q to a planar rectangle, $\phi(p_1) = 0$, $\phi(p_2) = 1$.

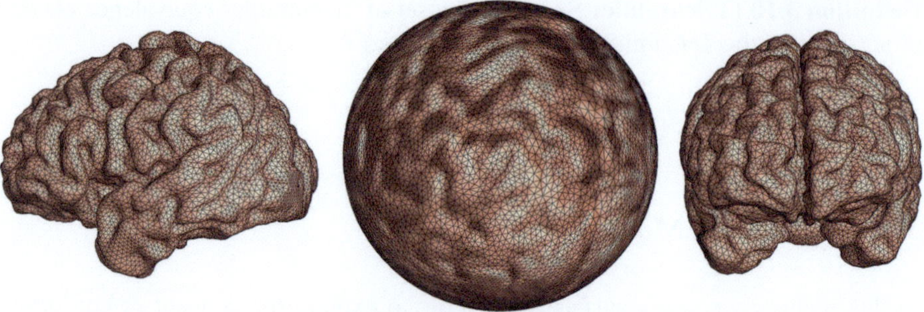

Fig. 3.4 Conformal mapping for a genus zero closed surface

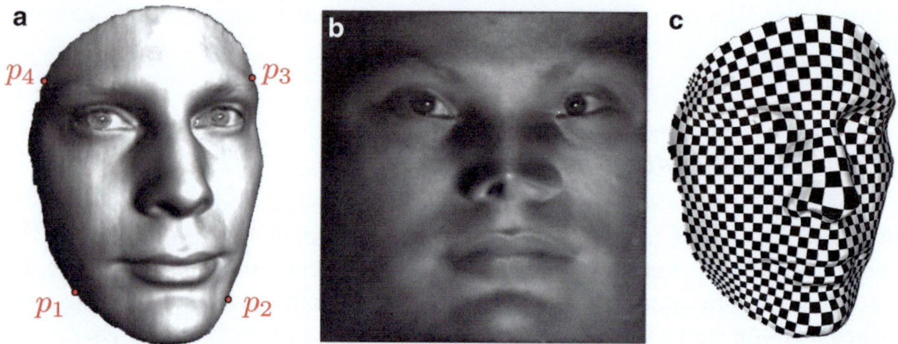

Fig. 3.5 Conformal module for a topological quadrilateral. The face surface with four boundary corners (**a**) is conformally mapped to a planar rectangle (**b**). A checkerboard texture is placed on the rectangle and pulled back to the face surface (**c**), all the right angles of checkers are well preserved

We can set the target Gauss curvature to be zero everywhere in the interior, and the target geodesic curvature to be zero along the boundary, except at the four corners $\{p_1, p_2, p_3, p_4\}$. The target exterior angle for those four corners is $\pi/2$. Then Ricci flow produces a flat metric. By isometrically embedding the surface with the flat metric onto the plane, we map it onto a planar rectangle, as shown in Fig. 3.5. The ratio between the height and the width of the rectangle is the conformal module of the topological quadrilateral. Therefore, the Teichmüller space for all topological quadrilaterals is one dimensional.

3.3.3   Topological Annulus

Suppose S is a topological annulus with a Riemannian metric $\mathbf{g}$. The boundary of S consists of 2 loops, $\partial S = \gamma_1 - \gamma_2$. There exists a conformal mapping $\phi : S \to \mathbb{C}$, which maps S to the canonical annulus, and $\phi(\gamma_1)$ and $\phi(\gamma_2)$ are concentric circles

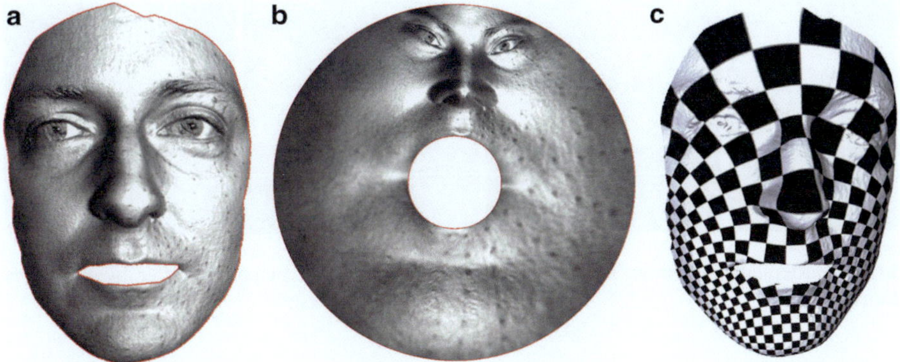

Fig. 3.6 Conformal module for a topological annulus. The face surface with two boundaries (**a**) is conformally mapped to a planar annulus (**b**). A checkerboard texture is placed on the annulus and pulled back to the face surface (**c**), all the right angles of checkers are well preserved

with radii r_1 and r_2 ($r_2 < r_1$), respectively, as shown in Fig. 3.6. The mapping ϕ is unique up to a planar rotation. The conformal module is defined as $\mathrm{Mod}(S) = \frac{1}{2\pi} \ln \frac{r_1}{r_2}$. Hence the Teichmüller space for all topological annuli is one dimensional, $dim T^{(0,2)} = 1$.

We can set the target Gauss curvature to be zero everywhere in the interior and the geodesic curvatures along the boundaries to be zero everywhere. Then Ricci flow leads to a flat metric. We find a curve γ connecting γ_1 and γ_2, such that γ is a straight line segment under the flat metric and orthogonal to the two boundaries. We slice the surface along γ to get $\tilde{S}$, and γ becomes two boundary segments γ^+ and γ^-. We then isometrically embed $\tilde{S}$ onto the plane. After a planar rigid motion, and a normalization, $\tilde{S}$ is a rectangle with the unit height, γ^- is on the real axis, and γ_1 is on the imaginary axis. Then we use the exponential map $\exp^{2\pi z}$ to map $\tilde{S}$ to a canonical planar annulus.

3.3.4 Topological Disk

Suppose S is a topological disk with a Riemannian metric. Then it can be conformally mapped to the unit planar disk. Two such mappings differ by a Möbius transformation

$$z \rightarrow e^{i\theta} \frac{z - z_0}{1 - \bar{z}_0 z}, \tag{3.1}$$

as shown in Fig. 3.7. The Teichmüller space of topological disks consists of a single point, $T^{(0,1)} = \{p\}$.

The computation is straightforward. We punch a small hole at the point z_0 to make the surface a topological annulus and map the annulus onto the canonical planar annulus. When the punched holes shrink to a point, the mappings converge to the Riemann mapping.

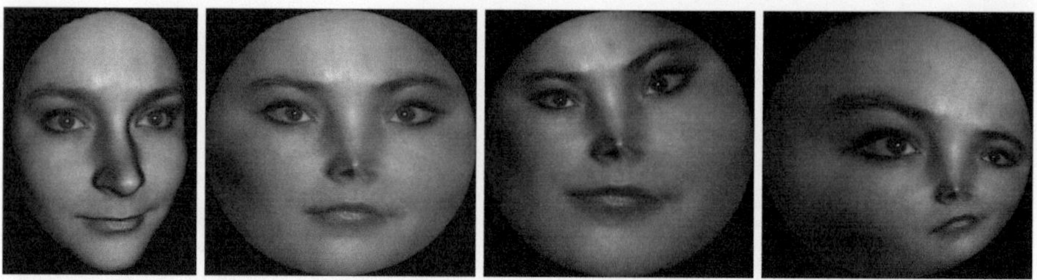

Fig. 3.7 Riemann mappings for a topological disk. Two such mappings differ by a Möbius transformation

3.3.5   *Topological Multiply Connected Annulus*

Suppose S is a genus zero surface with multiple boundaries. Then S is called a *topological multiply connected annulus*. Suppose S is with a Riemannian metric **g**. Then there exists a conformal mapping $\phi : S \to \mathbb{C}$, which maps S to the unit disk with circular holes (a planar circle domain). Such conformal mappings are unique up to Möbius transformations.

Let S be a multiply connected domain with $n + 1$ boundaries,

$$\partial S = \gamma_0 - \gamma_1 - \gamma_2 \cdots - \gamma_n, \tag{3.2}$$

where γ_0 is the exterior boundary and $\{\gamma_k, k > 0\}$ are sorted by their total lengths. We set the target Gauss curvature to be zeros for all the interior points, and the target geodesic curvatures to be constant on each boundary, such that

$$\int_{\gamma_0} \bar{k}_g = 2\pi, \quad \int_{\gamma_i} \bar{k}_g = -2\pi, \; i > 0.$$

Ricci flow produces a conformal mapping, which maps the surface to the planar circle domain. Let (x_k, y_k, r_k) represent the center and the radius of the circle image of $\gamma_k, k > 1$. We can use a Möbius transformation to map the center of γ_1 to the origin, and the center of γ_2 to be on the imaginary axis. Then the conformal module of the surface is given by

$$\{r_1, y_2, r_2, (x_3, y_3, r_3), \cdots, (x_n, y_n, r_n)\}.$$

So the dimension of the Teichmüller space of n-holed annuli is $3n - 3$, $dim T^{(0,n+1)} = 3n - 3$. Figure 3.8 shows the canonical conformal mapping of a multiply connected annulus, a partial surface of a 3D human face.

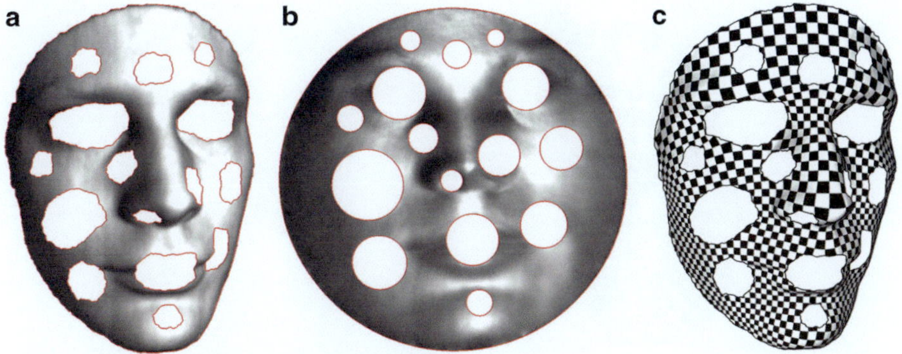

Fig. 3.8 Conformal module for a topological multiply connected annulus. The face surface with 16 boundaries (**a**) is conformally mapped to a planar circle domain (**b**). In the checkerboard texture mapping result (**c**), all the right angles of checkers are well preserved

3.3.6 *Topological Torus*

Given a topological torus (a genus one closed surface) S with a Riemannian metric, Ricci flow produces a flat metric. Then S is deformed to a flat torus T, and T can be represented as the quotient space,

$$T = \mathbb{C}/\Lambda,$$

where Λ is a lattice group,

$$\Lambda = \{z \to z + m + n\omega, m, n \in \mathbb{Z}\},\ \omega \in \mathbb{C},\ Img(\omega) > 0.$$

ω is the conformal module of the surface. Therefore, the Teichmüller space for genus one surfaces is a two-dimensional manifold, $dimT^{(1,0)} = 2$. Figure 3.9 shows the conformal module for a genus one closed surface.

3.3.7 *Genus One Surface with Boundaries*

We use the Ricci flow method to compute the canonical conformal mapping and the conformal module. Suppose the boundary of the surface has n loops, $\partial S = \gamma_1 + \gamma_2 + \cdots + \gamma_n$. We set the target Gauss curvature for all the interior points to be zero the target geodesic curvature on each boundary to be constant, and the total geodesic curvature for each boundary loop is -2π,

$$\int_{\gamma_k} \bar{k}_g = -2\pi,\ k = 1, 2, \cdots, n.$$

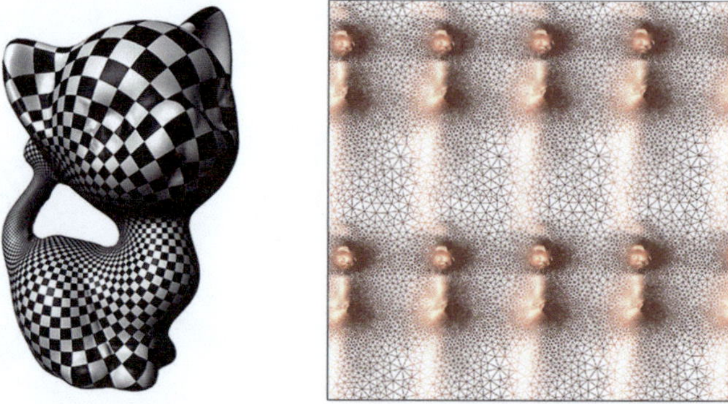

Fig. 3.9 Uniformization for genus one closed surface

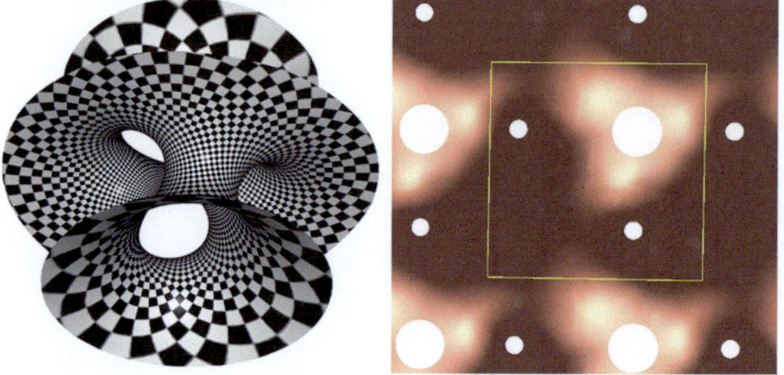

Fig. 3.10 Uniformization for genus one surface with boundaries, Costa's minimal surface

By running Ricci flow with the prescribed target curvature, a unique flat metric will be obtained. The surface is isometrically mapped onto a flat torus with circular holes,

$$\mathbb{C}/\Lambda - \bigcup_{k=1}^{n} D_k,$$

where $\{D_k\}$ are circles. The conformal module of the surface is given by the generators of the Lattice transformation group Λ and the centers and radii of the circular boundaries $\{D_k\}$. The dimension of the Teichmüller space is given by

$$dimT^{(1,n)} = 2 + 3n.$$

Figure 3.10 shows the conformal module for a genus one surface with three boundaries, Costa's minimal surface [2].

3.3.8 *High Genus Closed Surface*

Fuchs Group

Suppose S is a compact Riemann surface with genus $g > 1$. Let $q_0 \in S$ be the base point, select a set of canonical fundamental group generators $\{a_1, b_1, \cdots, a_g, b_g\}$, satisfying the following conditions

$$a_1 b_1 a_1^{-1} b_1^{-1} \cdots a_g b_g a_g^{-1} b_g^{-1} \sim e. \tag{3.3}$$

Suppose the universal covering of S is $p : \tilde{S} \to S$, where p is the projection. Let $z_0 \in \tilde{S}$ be one of the preimages of q_0, $z_0 \in p^{-1}(q_0)$. Through z_0, lift $a_1, b_1, a_2, b_2, \cdots, a_g, b_g$ to get paths,

$$\tilde{a}_1, \tilde{b}_1, \tilde{a}_2, \tilde{b}_2, \cdots, \tilde{a}_g, \tilde{b}_g.$$

Suppose the end points of these curves are

$$\xi_1, \eta_1, \xi_2, \eta_2, \cdots, \xi_g, \eta_g.$$

Then there must be $2g$ Deck transformations, $\alpha_1, \beta_1, \alpha_2, \beta_2, \cdots, \alpha_g, \beta_g$, such that

$$\alpha_j(z_0) = \xi_j, \beta_j(z_0) = \eta_j, \ j = 1, 2, \cdots, g.$$

Because of (3.3), we obtain

$$\alpha_1 \circ \beta_1 \circ \alpha_1^{-1} \circ \beta_1^{-1} \cdots \alpha_g \circ \beta_g \circ \alpha_g^{-1} \circ \beta_g^{-1} = id. \tag{3.4}$$

By Ricci flow, we can get the hyperbolic uniformization metric **g** of S, then the universal covering space $\tilde{S}$, with the pullback metric $p^*\mathbf{g}$, can be isometrically embedded in the hyperbolic space $\mathbb{H}^2$. The Deck transformations are hyperbolic isometric transformations, and called *Fuchsian transformations*, and the Deck transformation group is called the *Fuchs group* of the surface.

Poincaré Model

We use the Poincaré disk model for the hyperbolic plane

$$\mathbb{H}^2 = \{z| \ |z| < 1\}, \ ds^2 = \frac{4 dz d\bar{z}}{(1 - z\bar{z})^2}.$$

The hyperbolic lines through p and q are circular arcs passing through p and q, which are orthogonal to the unit circle. The hyperbolic circle $(\mathbf{c}, r)$ on the Poincaré disk is identical to the Euclidean circle $(\mathbf{C}, R)$, with the following relations,

$$C = \frac{1-t^2}{1-t^2 \mathbf{c}\bar{\mathbf{c}}}\mathbf{c}, \quad R = \sqrt{C\bar{C} - \frac{\mathbf{c}\bar{\mathbf{c}} - t^2}{1 - t^2 \mathbf{c}\bar{\mathbf{c}}}}, \quad t = \tanh\frac{r}{2}, \tag{3.5}$$

where $\mathbf{c}, \mathbf{C}$ are the hyperbolic and Euclidean circle centers, respectively, and r, R are the hyperbolic and Euclidean circle radii, respectively.

The Fuchsian transformations are Möbius transformations of hyperbolic type,

$$\phi(z) = e^{i\theta}\frac{z - z_0}{1 - \bar{z}_0 z}, \quad |z_0| < 1, \quad \theta \in [0, 2\pi),$$

which has two fixed points $\phi(\xi) = \xi$ and $\phi(\eta) = \eta$ on the unit circle. Let γ be a Möbius transformation for the extended complex plane, which maps the fixed points to 0 and ∞,

$$\gamma(z) = \frac{z - \xi}{z - \eta}.$$

Then

$$\gamma \circ \phi \circ \gamma^{-1} : z \to kz, \quad k \neq 0, 1,$$

where $k = \lambda$ is a positive real number, which is called *the multiplier* of the Möbius transformation. Therefore, a hyperbolic type Möbius transformation can be specified by its fixed points and multiplier (ξ, η, λ).

Fricke Coordinates

By (3.4), the Fuchs group generators α_g, β_g can be uniquely determined by α_1, $\beta_1, \cdots, \alpha_{g-1}, \beta_{g-1}$, which in turn can be represented by their fixed points and multipliers,

$$(\xi_1, \eta_1, \lambda_1, \cdots, \xi_{2g-2}, \eta_{2g-2}, \lambda_{2g-2}).$$

This is called the *Fricke coordinates* of the Riemann surface in the Teichmüller space T^g. So the Teichmüller space $T^g, g > 1$ is $6g - 6$ real dimensional, $dimT^g = 6g - 6$.

Figure 3.11 shows the isometric embedding of the universal covering spaces of two high genus surfaces on the Poincaré disk with their uniformization hyperbolic metrics.

Fig. 3.11 Uniformization for high genus surfaces

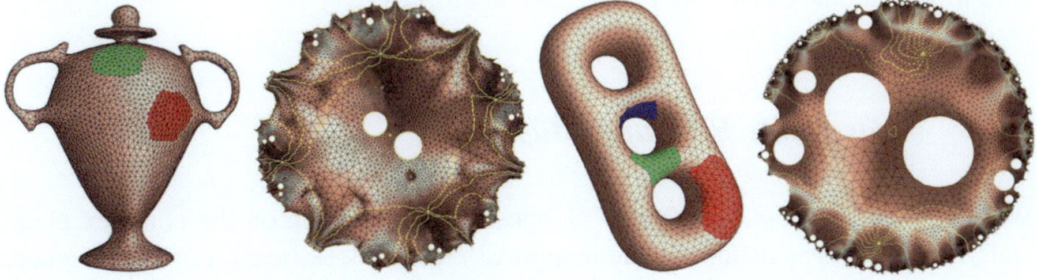

Fig. 3.12 Uniformization for high genus surfaces with multiple boundary components

3.3.9 *High Genus Surface with Boundaries*

A high genus surface S with n boundaries can be conformally mapped to a circle domain of a surface with the hyperbolic metric,

$$\mathbb{H}^2/\Lambda - \bigcup_{k=1}^{n} D_k,$$

where Λ is the Fuchs group of the surface, $\{D_k\}$ are hyperbolic circles. The Fricke coordinates of $\mathbb{H}^2/\Lambda$ and the centers and radii of $\{D_k\}$ form the conformal module of the surface. The dimension of the Techmüller space $T^{(g,n)}$ is $6g - 6 + 3n$, $dimT^{(g,n)} = 6g - g + 3n$.

Suppose the surface has boundaries $\partial S = \gamma_1 + \gamma_2 + \cdots + \gamma_n$. The target Gauss curvature is set to be -1 everywhere, the target geodesic curvatures on each γ_k is set to be a constant c_k, and the total geodesic curvature on γ_k is

$$\int_{\gamma_k} \bar{k}_g = -2\pi + \lambda_k,$$

where λ_k is chosen to ensure the trivial holonomy condition, namely, the Fuchsian transformation corresponding to γ_k must be identity, for $k = 1, 2, \cdots, n$. Details can be found in [14, 15]. Figure 3.12 demonstrates the canonical conformal mappings for high genus surfaces with multiple boundary components.

3.4 Quasi-Conformal Mapping

General diffeomorphisms between compact surfaces are rarely conformal, but are quasi-conformal. Geometrically, a conformal mapping maps infinitesimal circles to infinitesimal circles, whereas, a quasi-conformal mapping maps infinitesimal ellipses to infinitesimal circles. Figure 3.13 demonstrates both a conformal mapping and a quasi-conformal mapping from a 3D human face surface to the unit disk. We put the circle packing texture on the unit disk, and pull it back to the 3D face surface by the mapping. In the left two frames, the mapping is conformal, therefore, the circles are pulled back to circles; in the right two frames, the mapping is quasi-conformal, and the circles are pulled back to ellipses.

3.4.1 Measurable Riemann Mapping

We first consider planar mappings. Suppose D is a domain on the complex plane and $f : D \to \mathbb{C}$ is a differentiable mapping, $f : (x, y) \to (u, v)$. For convenience, we use complex representations: $z = x + iy$, $w = u + iv$, and $w = f(z)$. Then the mapping can be locally approximated by an affine transformation,

$$df = \frac{\partial f}{\partial z} dz + \frac{\partial f}{\partial \bar{z}} d\bar{z}. \tag{3.6}$$

If f is holomorphic, then $\partial f / \partial \bar{z} = 0$. The Jacobian of the mapping is given by

$$J = \left| \frac{\partial f}{\partial z} \right|^2 - \left| \frac{\partial f}{\partial \bar{z}} \right|^2.$$

If f is a diffeomorphism, then

$$\left| \frac{\partial f}{\partial \bar{z}} \right| < \left| \frac{\partial f}{\partial z} \right|. \tag{3.7}$$

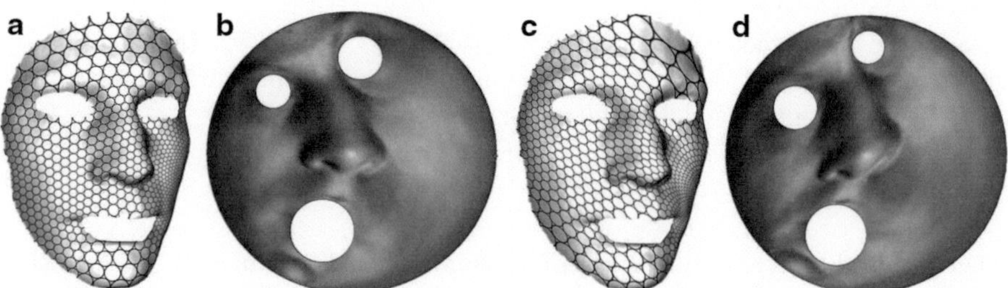

Fig. 3.13 Solving Beltrami equation using the auxiliary metric method. (**a–b**) A conformal mapping and its induced circle packing texture mapping for the $\mu \equiv 0$, (**c–d**) A quasi-conformal mapping and its induced circle packing texture mapping for a given μ

It follows from (3.6) that

$$(|f_z| - |f_{\bar{z}}|)|dz| \leq |dw| \leq (|f_z| + |f_{\bar{z}}|)|dz|,$$

where both limits can be attained. At the point z, the mapping maps an infinitesimal circle to an infinitesimal ellipse, and the ratio of the major axis to the minor axis is

$$K_f = \frac{|f_z| + |f_{\bar{z}}|}{|f_z| - |f_{\bar{z}}|} \geq 1.$$

This is called the *dilatation* at the point z. It is often more convenient to consider

$$k_f = \frac{|f_{\bar{z}}|}{|f_z|} < 1,$$

related to K_f by

$$K_f = \frac{1 + k_f}{1 - k_f}, \ k_f = \frac{K_f - 1}{K_f + 1}.$$

Definition 3.11 (Beltrami Coefficient). We introduce the *complex dilatation* or the *Beltrami coefficient* of the mapping f,

$$\mu_f = \frac{f_{\bar{z}}}{f_z},$$

with $|\mu_f| = k_f$.

The direction of the major axis of the ellipse corresponds to the direction

$$\arg dz = \alpha = \frac{1}{2} \arg \mu.$$

The minor axis direction is $\alpha \pm \pi/2$.

Definition 3.12 (Quasi-Conformal Mapping). The mapping f is said to be quasi-conformal if K_f is bounded. It is K-quasi-conformal if $K_f \leq K$.

The condition $K_f \leq K$ is equivalent to $k_f \leq (K_f - 1)/(K_f + 1)$. A 1-quasi-conformal mapping is conformal.

Suppose $f : D \to G$ and $g : D_1 \to G_1$ are two quasi-conformal mappings, and $G \subset D_1$. Then the Beltrami coefficient for the composition $g \circ f$ is given by

$$\mu_{g \circ f} = \frac{\mu_f + (\mu_g \circ f)\tau}{1 + \bar{\mu}_f \cdot (\mu_g \circ f)\tau}, \ \tau = \bar{f}_z/f_z.$$

Similarly, suppose $f : D \to G$ and $g : D \to G'$ are quasi-conformal mappings with Beltrami coefficients μ_f and μ_g, respectively. Then the Beltrami coefficient $g \circ f^{-1} : G \to G'$ is given by

$$\mu_{g \circ f^{-1}}(f) = \frac{\mu_g - \mu_f}{1 - \bar{\mu}_f \mu_g} \cdot \tau^{-1}, \quad \tau = \bar{f}_z / f_z.$$

If $f : \mathbb{D} \to \mathbb{D}$ is a quasi-conformal mapping with Beltrami coefficient μ_f, then $\|\mu_f\|_\infty < 1$; the inverse is also true.

Theorem 3.2 (Measurable Riemann Mapping). *Suppose Ω is a simply connected domain in $\mathbb{C}$, $\Omega \subset \mathbb{C}$, which has more than one boundary point. Assume $\mu(z)$ is a bounded measurable function defined in the interior of Ω and $\|\mu\|_\infty < 1$, then there exists a homeomorphism $f : \Omega \to \mathbb{D}$, which maps Ω to the unit disk, furthermore f satisfies the Beltrami equation*

$$\frac{\partial f}{\partial \bar{z}} = \mu(z) \frac{\partial f}{\partial z}. \tag{3.8}$$

Two such mappings $f_1, f_2 : \Omega \to \mathbb{D}$, differ by a Möbius transformation, namely, $f_2 \circ f_1^{-1} : \mathbb{D} \to \mathbb{D}$ has the following representation

$$f_2 \circ f_1^{-1}(z) = e^{i\theta} \frac{z - z_0}{1 - \bar{z}_0 z}, \quad |z_0| < 1, \ \theta \in [0, 2\pi).$$

Proof. We give a brief geometric proof based on the *Auxiliary Metric Method*. Let $f : \Omega \to \mathbb{D}$ be the quasi-conformal with the Beltrami coefficient μ, then

$$f : (\Omega, dz d\bar{z}) \to (\mathbb{D}, dw d\bar{w})$$

is quasi-conformal. Consider the *same* map f with a *different* Riemannian metric, and the pullback metric induced by f, $f^* dw d\bar{w} = dw d\bar{w}$, then

$$f : (\Omega, dw d\bar{w}) \to (\mathbb{D}, dw d\bar{w})$$

is isometric. The pullback metric

$$dw d\bar{w} = |w_z dz + w_{\bar{z}} d\bar{z}|^2 = |w_z|^2 \left| dz + \frac{w_{\bar{z}}}{w_z} d\bar{z} \right|^2 = |w_z|^2 |dz + \mu_f d\bar{z}|^2$$

is conformal to an auxiliary metric

$$\tilde{g} := |dz + \mu d\bar{z}|^2. \tag{3.9}$$

Then the same mapping f under the auxiliary metric

$$f : (\Omega, |dz + \mu d\bar{z}|^2) \to (\mathbb{D}, dw d\bar{w})$$

is a *conformal mapping*. According to Riemann mapping theorem, the conformal mapping f under the auxiliary metric exists and is unique up to a Möbius transformation. $\qquad\Box$

Although the proof is for topological disk, it can be directly generalized to compact Riemann surfaces with arbitrary topologies, where the concept of Beltrami coefficient $\mu_f = f_{\bar{z}}/f_z$ needs to be replaced by the concept of Beltrami differential.

Definition 3.13 (Beltrami Differential). Suppose $f : S_1 \to S_2$ is a diffeomorphism between two Riemann surfaces S_1 and S_2. Suppose z and w are local conformal parameters of S_1 and S_2, respectively, and f has the local representation $w = f(z)$. Then the Beltrami differential of f has the local representation

$$\mu_f := \frac{f_{\bar{z}}}{f_z} \frac{d\bar{z}}{dz}.$$

We can solve Beltrami equations (3.8) on surfaces with complicated topologies using Ricci flow combined with the auxiliary metric method. Figure 3.13 shows the computational results for multiply connected annulus surfaces.

3.4.2 Existence of Isothermal Coordinates

The measurable Riemann mapping theorem leads to the existence of isothermal coordinates.

Corollary 3.1 (Isothermal Coordinates). *Let S be a smooth metric surface. For any point $p \in S$, there exists a neighborhood $U(p)$. The isothermal coordinates exist on $U(p)$.*

Proof. Suppose S is with a Riemannian metric

$$g(x, y) = \begin{pmatrix} g_{11} & g_{12} \\ g_{21} & g_{22} \end{pmatrix} = \begin{pmatrix} \cos\theta & \sin\theta \\ -\sin\theta & \cos\theta \end{pmatrix} \begin{pmatrix} \lambda_1 & 0 \\ 0 & \lambda_2 \end{pmatrix} \begin{pmatrix} \cos\theta & -\sin\theta \\ \sin\theta & \cos\theta \end{pmatrix}, \lambda_1 > \lambda_2 > 0.$$

Choose a neighborhood $U(p)$, define Beltrami coefficient

$$\mu(z) = \frac{\lambda_1 - \lambda_2}{\lambda_1 + \lambda_2} e^{i2\theta}, \quad z = x + iy,$$

and solve the Beltrami equation $w_{\bar{z}} = \mu(z)w_z$, where $w = u + iv$. Because $\|\mu\|_\infty < 1$, the equation has a solution ($\|\mu\|_\infty$ is the maximum of the norm of μ on S). Then (u, v) are the isothermal coordinates of the surface. $\qquad\Box$

3.4.3 Conformal Surface Representation

In order to represent a parametric surface embedded in $\mathbb{E}^3$, generally, we need three functions, $\mathbf{r}(u, v) = (x(u, v), y(u, v), z(u, v))$, to represent its position. In fact, by using isothermal coordinates, only two functions are sufficient.

Theorem 3.3. *A closed surface embedded in $\mathbb{E}^3$ with isothermal coordinates (u, v) is determined by the conformal factor $\lambda(u, v)$ and the mean curvature $H(u, v)$ uniquely up to a rigid motion. A simply connected surface with a boundary is determined by the conformal factor, the mean curvature and the boundary position.*

Proof. Because $z = u + iv$ is isothermal, we have

$$\lambda^2 = \langle \mathbf{r}_u, \mathbf{r}_u \rangle = \langle \mathbf{r}_v, \mathbf{r}_v \rangle,$$

$$\langle \mathbf{r}_u, \mathbf{r}_v \rangle = 0.$$

This is equivalent to

$$\langle \mathbf{r}_z, \mathbf{r}_z \rangle = 0, \ \langle \mathbf{r}_{\bar{z}}, \mathbf{r}_{\bar{z}} \rangle = 0, \ \langle \mathbf{r}_z, \mathbf{r}_{\bar{z}} \rangle = \frac{\lambda^2}{2}.$$

The natural frame of the surface is given by $\{\mathbf{r}_z, \mathbf{r}_{\bar{z}}, \mathbf{n}\}$, where $\mathbf{n}$ is the normal field of the surface. Let

$$\mu = \langle \mathbf{r}_{zz}, \mathbf{n} \rangle,$$

then the motion equation for the natural frame can be formulated as

$$\begin{aligned}
\mathbf{r}_{zz} &= \tfrac{2}{\lambda}\lambda_z \mathbf{r}_z + \mu \mathbf{n}, \\
\mathbf{r}_{z\bar{z}} &= \tfrac{\lambda^2}{2} H \mathbf{n}, \\
\mathbf{n}_z &= -H \mathbf{r}_z - \tfrac{2}{\lambda^2} \mu \mathbf{r}_{\bar{z}},
\end{aligned} \tag{3.10}$$

with Gauss equation (3.11) and Codazzi equation (3.12)

$$(\log \lambda)_{z\bar{z}} = \frac{\mu \bar{\mu}}{\lambda^2} - \frac{\lambda^2}{4} H^2, \tag{3.11}$$

$$\mu_{\bar{z}} = \frac{\lambda^2}{2} H_z. \tag{3.12}$$

The Codazzi equation is equivalent to the following Poisson problem

$$\mu_{z\bar{z}} = \frac{1}{2}\lambda(2\lambda_z H_z + \lambda H_{zz}).$$

One can solve μ from this Poisson equation, then integrate the natural frame $\{\mathbf{r}_z, \mathbf{r}_{\bar{z}}, \mathbf{n}\}$ to reconstruct the surface. This completes the proof. $\square$

Corollary 3.2. *Given a closed surface embedded in $\mathbb{E}^3$ with isothermal coordinates (conformal structure), if the two principle curvatures $k_1(u, v)$ and $k_2(u, v)$ are known, then the surface can be determined uniquely up to a rigid motion of $\mathbb{E}^3$.*

Proof. From the Gauss curvature $K = k_1 k_2$ and the conformal structure, one can compute the conformal factor λ using Ricci flow. Combined with the mean curvature $H = (k_1 + k_2)/2$, the surface can be determined uniquely up to a rigid motion. $\square$

3.4.4 Diffeomorphism Space and Beltrami Holomorphic Flow

Suppose metric surfaces $(S_1, \mathbf{g}_1)$ and $(S_2, \mathbf{g}_2)$ are two topological disks. Consider the space of all diffeomorphisms between them and the space of Beltrami coefficients. From measurable Riemann mapping theorem, we obtain the following relation

$$\frac{\{Diffeomorphisms\}}{\{M\ddot{o}bius\}} \cong \{Beltrami\ Coefficients\}.$$

In what follows, f^μ will always denote the solution to the Beltrami equation on the extended complex plane $\bar{\mathbb{C}} = \mathbb{C} \cup \{\infty\}$ with fixed points $0, 1, \infty$. Let

$$R(z, \zeta) = \frac{1}{z - \zeta} - \frac{\zeta}{1 - z} + \frac{\zeta - 1}{z}.$$

Suppose

$$\mu(t + s)(z) = \mu(t)(z) + s\nu(t)(z) + s\epsilon(s, t)(z),$$

where

$$\nu(t), \mu(t), \epsilon(s, t) \in L^\infty, \ \|\mu(t)\|_\infty < 1,$$

and $\|\epsilon(s, t)\|_\infty \to 0$ as $s \to 0$, where $\|\epsilon(s, t)\|_\infty$ is defined as $\sup_{z \in \bar{\mathbb{C}}} \|\epsilon(s, t)(z)\|$. Then

$$f^{\mu(t+s)}(\zeta) = f^{\mu(t)}(\zeta) + s\dot{f}(\zeta, t) + o(s)$$

uniformly on compact sets, where

$$\dot{f}(\zeta, t) = -\frac{1}{\pi} \int \int \nu(t)(z) R(f^{\mu(t)}(z), f^{\mu(t)}(\zeta))(f_z^{\mu(t)}(z))^2 dx dy. \tag{3.13}$$

This formula gives the dependence of the mapping on the Beltrami coefficients. Besides the auxiliary metric method, this variation formula gives a different way to solve Beltrami equations $f_{\bar{z}} = \mu f_z$. We define a homotopy $\mu(t) = t\mu, t \in [0, 1]$. Obviously, $f^{\mu(0)}$ is a conformal mapping. Then we use the above formula to deform the mapping from $f^{\mu(0)}$ to $f^{\mu(1)}$. This method is called the *Beltrami holomorphic flow* method [8].

Furthermore, this formula can be applied for many optimization problems in the diffeomorphism space.

3.4.5   *Teichmüller Map and Teichmüller Distance*

Let $\phi(z)dz^2$ be a holomorphic quadratic differential on a Riemann surface S. Let p be a regular point, such that $\phi(p) \neq 0$. Then we can choose a neighborhood $U(p) = \{z, |z - p| < r\}$ of p, such that ϕ is nonzero inside $U(p)$. Then choose one branch of $\sqrt{\phi(z)}$, and define

$$\zeta(z) = \int_0^z \sqrt{\phi(t)}dt,$$

the mapping $z \to \zeta$ is biholomorphic, and ζ is called the natural parameter induced by ϕ.

Definition 3.14 (Teichmüller Map). Let S_0 and S_1 be two Riemann surfaces and $f : S_0 \to S_1$ be a quasi-conformal map. Suppose ω_0 and ω_1 are holomorphic quadratic differentials on S_0 and S_1, respectively. If f maps the regular points of ω_0 to the regular points of ω_1 and the zeros of ω_0 to the zeros of ω_1 with the same orders, and let ζ_0 and ζ_1 be the natural parameters induced by ω_0 and ω_1, f has the local representation

$$f : \zeta_0 \to \zeta_1 = \frac{K+1}{2}\zeta_0 + \frac{K-1}{2}\bar{\zeta}_0,$$

where $K \geq 1$ is a constant, then we call f a Teichmüller map, ω_0 the source holomorphic quadratic differential, and ω_1 the target holomorphic quadratic differential.

The followings are the fundamental theorems in Teichmüller theory.

Theorem 3.4 (Existence of Teichmüller Map). *Suppose S_0 and S_1 are two Riemann surfaces with genus $g > 1$. If $f : S_0 \to S_1$ is an orientation preserving homeomorphism, then there exists a Teichmüller map homotopic to f.*

Theorem 3.5 (Uniqueness of Teichmüller Map). *Suppose S_0 and S_1 are two Riemann surfaces with genus $g > 1$. If $f_0 : S_0 \to S_1$ is a Teichmüller map, then for any map $f : S_0 \to S_1$ homotopic to f_0,*

$$K[f] \geq K[f_0],$$

and the equality holds if and only if $f = f_0$.

Teichmüller map gives a Riemannian metric for the Teichmüller space.

Definition 3.15 (Teichmüller Distance). Suppose S is a topological surface, and μ_1 and μ_2 are two conformal structures on S. $S_{\mu_1}, S_{\mu_2} \in T^g$ are two Riemann surfaces. The Teichmüller distance between the two Riemann surfaces is given by

$$d_T(S_{\mu_1}, S_{\mu_2}) = \inf_{f \sim id} \log K[f].$$

Then the dilatation of Teichmüller map between S_{μ_1} and S_{μ_2} gives the distance. The Teichmüller space under the Teichmüller distance is a Riemannian manifold.

Theorem 3.6. *The Teichmüller space T^g, $g \geq 1$ with the Teichmüller distance d_T is a Riemannian manifold (T^g, d_T).*

3.5 Harmonic Maps

Harmonic maps are the natural generalizations of geodesics, which have been broadly applied for computing the diffeomorphisms among surfaces. In the following discussion, we always use isothermal coordinates, unless otherwise specified.

Suppose a map $w : (M, \sigma|dz|^2) \to (N, \rho|dw|^2)$ is given. If for any two points $p, q \in M$,

$$d_\rho(w(p), w(q)) < K d_\sigma(p, q),$$

where K is a positive constant, then the map is a Lipschitz map. The *energy density* at a point z is defined to be

$$e(w; \sigma, \rho) = \frac{\rho(w(z))}{\sigma(z)}|w_z|^2 + \frac{\rho(w(z))}{\sigma(z)}|w_{\bar{z}}|^2.$$

The *harmonic energy* of the map is defined to be

$$E(w; \sigma, \rho) = \int_M e(w; \sigma, \rho)\sigma \, dz d\bar{z}.$$

The total energy depends on the metric of the target surface and the conformal structure of the source surface.

Definition 3.16 (Harmonic Map). A Lipschitz map $w : (M, \sigma|dz|^2) \to (N, \rho|dw|^2)$ is called a harmonic map if it is a critical point of the energy functional $E(w; \sigma, \rho)$ in the mapping space.

The Euler–Lagrange equation for the energy functional is

$$w_{z\bar{z}} + (\log \rho)_w w_z w_{\bar{z}} = 0. \tag{3.14}$$

Given an initial mapping, the *heat flow method*,

$$\frac{\partial w}{\partial t} = -w_{z\bar{z}} - (\log \rho)_w w_z w_{\bar{z}},$$

can deform it to a harmonic map.

3.5.1   Topological Disk

Suppose a harmonic mapping is from a topological disk to a planar disk. If the restriction on the boundary is a homeomorphism, then the map is a diffeomorphism.

Theorem 3.7 (Radó [11]). *Assume $\Omega \subset \mathbb{E}^2$ is a convex domain with a smooth boundary $\partial\Omega$ and a metric surface $(S, \mathbf{g})$ is a simply connected domain with a single boundary. Given any homeomorphism $\phi : \partial S \to \partial\Omega$, then the harmonic map $u : S \to \Omega$, such that $u = \phi$ on ∂S, is a diffeomorphism.*

Proof. We can map S conformally onto the planar disk $\mathbb{D}$ by a Riemann mapping $\tau : S \to \mathbb{D}$, then the composition $\phi \circ \tau^{-1} : \mathbb{D} \to \Omega$ is still harmonic. We denote the map as $w(z)$, or in real form $(u(x, y), v(x, y))$. Then both u and v are harmonic functions,

$$\Delta u \equiv 0, \ \Delta v \equiv 0.$$

Suppose at one interior point p, the Jacobian is degenerated. Then there are two real numbers $(\lambda_1, \lambda_2) \neq (0, 0)$, such that

$$\lambda_1 \nabla u + \lambda_2 \nabla v = \mathbf{0}.$$

Define a linear function $f : \Omega \to \mathbb{R}$, $f(u, v) = \lambda_1 u + \lambda_2 v$, then

$$f(x, y) = \lambda_1 u(x, y) + \lambda_2 v(x, y)$$

is a harmonic function defined on $\mathbb{D}$, with an interior singularity p. Because f is harmonic, p cannot be local maximum or minimum, then p must be a saddle point. Therefore the level set $f = f(p) + \epsilon$ has two branches and intersects the boundary of $\mathbb{D}$ at four points $\{z_1, z_2, z_3, z_4\}$. But each line $\lambda_1 u + \lambda_2 v = const$ has only two intersection points $\{w_1, w_2\}$ with $\partial\Omega$, because Ω is convex. Therefore, the mapping $w(z)$ maps four points $\{z_i\}$ to the two points $\{w_j\}$, so the mapping ϕ restricted on the boundary ∂S to $\partial\Omega$ is not a homeomorphism. This contradicts the assumption that $\phi : \partial S \to \partial\Omega$ is a homeomorphsim. Therefore, the Jacobian is nonzero everywhere in the interior of the surface, namely, the mapping is a diffeomorphism. $\square$

Furthermore, the harmonic map is unique.

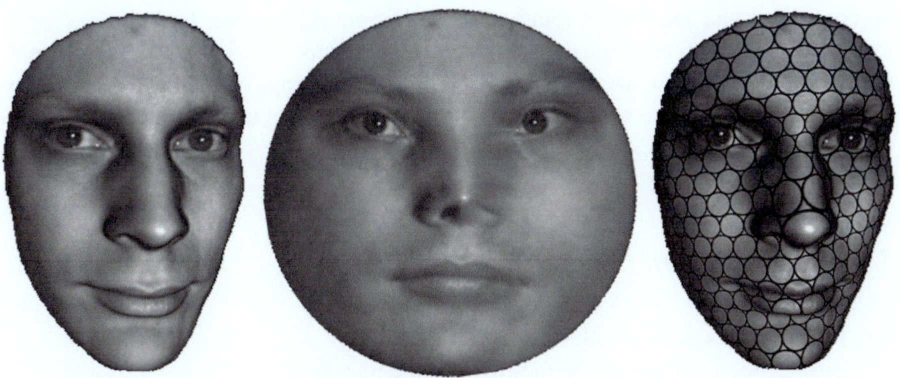

Fig. 3.14 Harmonic map for a human face surface

Theorem 3.8. *Assume* $\Omega \subset \mathbb{E}^2$ *is a simply connected domain with a smooth boundary* $\partial\Omega$ *and a metric surface* $(S, \mathbf{g})$ *is simply connected with a single boundary. Given any homeomorphism* $\phi : \partial S \rightarrow \partial\Omega$, *if there exists a harmonic map* $u : S \rightarrow \Omega$, *such that* $u = \phi$ *on* ∂S, *then* u *is unique.*

Proof. The uniqueness of harmonic maps between topological disks can be easily shown using the *mean value* property. By using Riemann mapping, the harmonic mapping between two metric surfaces which are topological disks is converted to a harmonic mapping from the unit disk to itself, $w(z) = (u(x, y), v(x, y))$, where u, v are harmonic functions

$$\begin{cases} w_{z\bar{z}} \equiv 0 \\ w|_{\partial\mathbb{D}} = w_0 \end{cases}.$$

From Poisson's formula

$$w(re^{i\phi}) = \frac{1}{2\pi} \int_0^{2\pi} w_0(e^{i\theta}) \frac{1 - r^2}{1 - 2r\cos(\theta - \phi) + r^2} d\theta,$$

we get the mean value property. Namely, for any interior point $z_0 \in \mathbb{D}$, one can choose a small circle surrounding z_0, $z_0 + \epsilon e^{i\theta}$,

$$w(z_0) = \frac{1}{2\pi\epsilon} \int_0^{2\pi} w(z_0 + \epsilon e^{i\theta}) d\theta,$$

therefore, the maximum and the minimum of u and v must be on the boundary of the unit disk.

Suppose there are two harmonic mappings $w_1, w_2 : \mathbb{D} \rightarrow \mathbb{D}$ with the same boundary conditions $w_0 : \partial\mathbb{D} \rightarrow \partial\mathbb{D}$. Then the maximum and the minimum of the harmonic function $u_1 - u_2$ are on the boundary, therefore 0's, so $u_1 \equiv u_2$. Similarly, $v_1 \equiv v_2$, therefore $w_1 \equiv w_2$. $\qquad\square$

Figure 3.14 shows an example for a harmonic map from a human face surface onto the planar disk. From example, one can see that the harmonic map is close to be conformal and introduces relatively small angle distortions. In fact, conformal mappings are harmonic, but harmonic maps may not necessarily be conformal. In the figure, if we only fix the images of 3 boundary points, and allow the boundary images $\phi(p)$, $p \in \partial S$ to slide along ∂D, and further minimize harmonic energy, then when the energy reaches the minimum, we obtain a unique conformal mapping.

3.5.2 Genus Zero Closed Surface

The following discussion follows the proof in [9].

Consider the pullback of the metric ρ by the map w,

$$w^*\rho = \rho\,dw\,d\bar{w} = \rho w_z \overline{w_{\bar{z}}}\,dz^2 + (\rho|w_z|^2 + \rho|w_{\bar{z}}|^2)\,dz\,d\bar{z} + \overline{\rho w_z \overline{w_{\bar{z}}}}\,d\bar{z}^2.$$

The Hopf differential is defined to be the $(2, 0)$ part of this pullback,

$$\Phi(w) = \rho w_z \overline{w_{\bar{z}}}\,dz^2.$$

When w is harmonic, the Hopf differential $\Phi(w)$ is holomorphic. If w is conformal, then $\Phi(w)$ vanishes. Hence we get the following,

Proposition 3.1. *Suppose w is an orientation preserving diffeomorphism. If w is harmonic, then the Hopf differential $\Phi(w)$ is holomorphic. If w is conformal, then $\Phi(w)$ vanishes.*

Proof. Take the partial derivative of Hopf differential $\phi(z)$,

$$\phi(z)_{\bar{z}} = (\rho_w w_{\bar{z}} w_z + \rho w_{z\bar{z}})\overline{w_{\bar{z}}} + (\rho_{\bar{w}}(\bar{w})_{\bar{z}}\overline{w_{\bar{z}}} + \rho(\overline{w_{\bar{z}}})_{\bar{z}})w_z.$$

The first term is zero, due to (3.14). Because $\overline{w_{\bar{z}}} = (\bar{w})_z$, the second term is

$$(\rho_{\bar{w}}(\bar{w})_{\bar{z}}(\bar{w})_z + \rho(\bar{w})_{z\bar{z}})w_z,$$

which is also zero. Therefore, if w is harmonic, then its Hopf differential is a holomorphic quadratic differential. If w is an orientation preserving diffeomorphism, then its Jacobian is $|w_z|^2 - |w_{\bar{w}}|^2 > 0$. $\phi(z) = \rho w_z w_{\bar{z}}$ vanishes if and only if $w_{\bar{z}} = 0$, namely, w is conformal. $\qquad\square$

Theorem 3.9. *Harmonic maps between genus zero closed metric surfaces are conformal.*

Proof. Suppose w is harmonic. Then its Hopf differential $\Phi(w)$ is a holomorphic quadratic differential. On genus zero closed metric surfaces, holomorphic quadratic differentials vanish, therefore, w is conformal. $\qquad\square$

3.5.3 *High Genus Closed Surface*

In the following, we show that if the target surface has negative curvature everywhere, then degree one harmonic map is diffeomorphic. We follow the proof in [9]. More details can also be found in [10, 11].

We can define the auxiliary functions,

$$
\begin{aligned}
\mathscr{H} &= \frac{\rho(w(z))}{\sigma(z)}|w_z|^2, \\
\mathscr{L} &= \frac{\rho(w(z))}{\sigma(z)}|w_{\bar{z}}|^2.
\end{aligned}
$$

The Jacobian of the mapping $\mathscr{J}$ is given by

$$
\mathscr{J} = \mathscr{H} - \mathscr{L}.
$$

The complex form of the Laplace–Beltrami operator on $(M, \sigma|dz|^2)$ is

$$
\Delta_\sigma = \frac{4}{\sigma}\frac{\partial^2}{\partial z \partial \bar{z}},
$$

then the Gauss curvatures are given by

$$
K(\sigma) = -\frac{1}{2}\Delta_\sigma \log \sigma, \quad K(\rho) = -\frac{1}{2}\Delta_\rho \log \rho.
$$

This gives the Bochner formula,

$$
\begin{aligned}
\Delta_\sigma \log \mathscr{H} &= -2K(\rho)\mathscr{H} + 2K(\rho)\mathscr{L} + 2K(\sigma) = -2K(\rho)\mathscr{J} + 2K(\sigma), \\
\Delta_\sigma \log \mathscr{L} &= -2K(\rho)\mathscr{L} + 2K(\rho)\mathscr{H} + 2K(\sigma) = 2K(\rho)\mathscr{J} + 2K(\sigma).
\end{aligned}
$$

The difference between these two equations gives the relation between the Jacobian of the map and the Gauss curvature on the target surface,

$$
\Delta_\sigma \log \frac{\mathscr{H}}{\mathscr{L}} = -4K(\rho)\mathscr{J}.
$$

In case M and N are compact surfaces of genus g_M and g_N, respectively, $w : M \to N$ is a harmonic map of degree $deg(w)$, then the function $\mathscr{H}$ either vanishes identically or has at most isolated zeros. Each zero point $p \in M$ has an order $n_p \in \mathbb{Z}, n_p \geq 0$, and the summation of the orders of all zero points is

$$
\sum_{\mathscr{H}(p)=0} n_p = -deg(w)(4g_N - 4) + (4g_M - 4). \tag{3.15}
$$

Similarly, if $\mathscr{L}$ is not identically zero, then

$$
\sum_{\mathscr{L}(p)=0} n_p = deg(w)(4g_N - 4) + (4g_M - 4).
$$

Suppose $\mathscr{J}$ is not identically zero, and $p \in M$ is a zero point of $\mathscr{J}$, $\mathscr{J}(p) = 0$, then one can show $\mathscr{H}(p) = 0$ and $\mathscr{L}(p) = 0$. Therefore, the zero points of $\mathscr{J}$ must be isolated. If $\mathscr{J} \geq 0$ everywhere, and the number of preimages $\{w^{-1}(q)\}$ of a regular value $q \in N$ is bounded, then the zero of $\mathscr{J}$ is a nontrivial branch point of w.

Theorem 3.10. *If the metric surface $(M, \sigma|dz|^2)$ and $(N, \rho|dw|^2)$ have the same genus $g \geq 1$, and the Gauss curvature $K(\rho)$ on the target is nonpositive everywhere, $K(\rho) \leq 0$, then a degree 1 harmonic map $w : M \to N$ must be a diffeomorphism.*

Proof. First, we claim $\mathscr{J} \geq 0$ everywhere. Otherwise, assume at a point $p \in M$, $\mathscr{J} < 0$. We can define a region $D \subset M$,

$$D = \{q \in M \,|\, \mathscr{J}(q) < 0\},$$

Because $p \in D$, D is nonempty. In the region $D \cup \partial D$, $\mathscr{J} \leq 0$,

$$\Delta_\sigma \log \frac{\mathscr{H}}{\mathscr{L}} = -2K(\rho)\mathscr{J} \leq 0,$$

so $\log \frac{\mathscr{H}}{\mathscr{L}}$ is superharmonic. From formula (3.15), the total order of zeros of $\mathscr{H}$ is

$$\sum_{\mathscr{H}(p)=0} n_p = 0,$$

therefore $\mathscr{H}$ has no zeros, and $\mathscr{H} > 0$ everywhere. On the boundary of D, $p \in \partial D$,

$$\mathscr{J}(p) = \mathscr{H} - \mathscr{L} = 0, \quad \log \frac{\mathscr{H}}{\mathscr{L}} = 0,$$

so $\mathscr{L}(p) > 0$. For any interior point $q \in D$, $\log \frac{\mathscr{H}}{\mathscr{L}}$ is no less than the boundary value 0, therefore

$$\log \frac{\mathscr{H}}{\mathscr{L}} \geq 0, \quad \mathscr{H} \geq \mathscr{L},$$

hence $\mathscr{J} = \mathscr{H} - \mathscr{L} \geq 0$ in the whole D. This contradicts the assumption $\mathscr{J}(p) < 0$, $p \in D$. Therefore, $\mathscr{J} \geq 0$ everywhere.

From the above discussion, $\mathscr{J}$ has at most isolated zeros which are nontrivial branch points. Since the degree of the map is 1, these zeros do not exist, so we have $\mathscr{J} > 0$ on M and w is a diffeomorphism. $\qquad\square$

Furthermore, if the target surface is of genus $g > 1$ and the Gauss curvature is nonpositive, then the degree one harmonic map exists [6] and is unique [5].

3.5.4 Teichmüller Space Representation

Finally, we show that the space of all holomorphic quadratic differentials Φdz^2 on $(M, \sigma |dz|^2)$, denoted as $QD(\sigma)$, has one-to-one correspondence to the Teichmüller space T^g, where T^g is represented as all possible hyperbolic metrics on N, $(N, \rho |dw|^2)$.

Theorem 3.11. *Suppose $w : (M, \sigma |dz|^2) \rightarrow (N, \rho |dw|^2)$ where $K(\sigma) \equiv -1$, $K(\rho) \equiv -1$, and w is harmonic. The Hopf differential Φ induces a map from the space of holomorphic quadratic differentials $QD(\sigma)$ to the Teichmüller space T^g. Φ is injective.*

Proof. Suppose there are two harmonic maps w_1, w_2 from $(M, \sigma |dz|^2)$ to $(N, \rho_1 |dw|^2)$ and $(N, \rho_2 |dw|^2)$, respectively, where $K(\rho_1) \equiv -1$ and $K(\rho_2) \equiv -1$, and they induce the same Hopf differential $\Phi_1 = \Phi_2 = \Phi$. Then for $h = \log \mathcal{H}$,

$$\Delta h = 2e^h - 2|\Phi|^2 e^{-h} - 2.$$

If $h_1 > h_2$ somewhere, then we can look at a maximum of $h_1 - h_2$ and find that

$$0 \geq \Delta(h_1 - h_2) = (e^{h_1} - e^{h_2}) - |\Phi|^2(e^{-h_1} - e^{-h_2}) > 0,$$

so $h_1 \leq h_2$. Symmetrically, $h_2 \leq h_1$, so $h_1 = h_2$ everywhere. Then $\mathcal{H}_1 = \mathcal{H}_2$. Since $\mathcal{H}_1 \mathcal{L}_1 = |\Phi|^2 = \mathcal{H}_2 \mathcal{L}_2$ and $\mathcal{H}_k > 0$, we obtain $\mathcal{L}_1 = \mathcal{L}_2$. Consequently, the densities $e_k = \mathcal{H}_k + \mathcal{L}_k, k = 1, 2$ are equal. Therefore, the pullback metrics

$$w_k^* \rho = \Phi_k dz^2 + \sigma e_k dz d\bar{z} + \bar{\Phi}_k d\bar{z}^2, \; k = 1, 2,$$

are equal so that $w_1 \circ w_2^{-1} : (N, \rho_2) \rightarrow (N, \rho_1)$ is an isometry isotopic to the identity. $\square$

Therefore, holomorphic quadratic differentials can be treated as the tangent space of the Teichmüller space.

References

1. Ahlfors, L.V.: Lectures on Quasiconformal Mappings. University Lecture Series, vol. 38, 2nd edn. American Mathematical Society, Providence, RI (2006). With supplemental chapters by C. J. Earle, I. Kra, M. Shishikura and J. H. Hubbard
2. Costa, C.: Example of a complete minimal immersion in $\mathbb{R}^3$ of genus one and three embedded ends. Bull. Soc. Bras. Mat. **15**, 47–54 (1984)
3. Gardiner, F., Lakic, N.: Quasiconformal Teichmüler Theory. American Mathematical Society, Providence, RI (1999)
4. Gu, X.D., Yau, S.T.: Computational Conformal Geometry. Advanced Lectures in Mathematics. High Education Press and International Press, Somerville (2008)

5. Hartman, P.: On homotopic harmonic maps. Can. Math. Soc. **19**, 673–687 (1967)
6. James Eells, J., Sampson, J.H.: Harmonic mappings of riemannian manifolds. Am. J. Math. **86**(1), 109–160 (1964)
7. Jost, J.: Compact Riemann Surfaces: An Introduction to Contemporary Mathematics. Universitext. Springer, Berlin (1997)
8. Lui, L.M., Wong, T.W., Zeng, W., Gu, X., Thompson, P.M., Chan, T.F., Yau, S.T.: Optimization of surface registrations using Beltrami holomorphic flow. J. Sci. Comput. **50**(3), 557–585 (2012)
9. Schoen, R., Yau, S.T.: On univalent harmonic maps between surfaces. Invent. Math. **44**(3), 265–278 (1978)
10. Schoen, R., Yau, S.T.: Lecture on Differential Geometry, vol. 1. International Press Incorporated, Boston (1994)
11. Schoen, R., Yau, S.T.: Lecture on Harmonic Maps, vol. 2. International Press Incorporated, Boston (1997)
12. Seppala, M., Sorvali, T.: Geometry of Riemann Surfaces and Teichmüller Spaces. North-Holland Mathematics Studies. North-Holland Publishing, Amsterdam (1992)
13. Wolf, M.: The Teichmüller theory of harmonic maps. J. Differ. Geom. **29**(2), 449–479 (1989)
14. Zeng, W., Yin, X., Zhang, M., Luo, F., Gu, X.: Generalized Koebe's method for conformal mapping multiply connected domains. In: ACM Symposium on Solid and Physical Modeling (SPM'09), pp. 89–100, ACM, New York, NY, 2009
15. Zhang, M., Li, Y., Zeng, W., Yau, S.T., Gu, X.: Canonical conformal mapping for high genus surfaces with boundaries. Int. J. Comput. Graph. **36**(5), 417–426 (2012)

Chapter 4
Discrete Surface Ricci Flow

This chapter systematically introduces the discrete surface Ricci flow theory. The whole theory is explained using the variational principle on discrete surfaces based on derivative cosine law [18].

Ricci flow conformally deforms the Riemannian metrics, such that during the flow the infinitesimal circles are preserved. This phenomenon inspired Thurston to develop the circle packing method. In his work on constructing hyperbolic metrics on 3-manifolds, Thurston [25] studied a Euclidean (or a hyperbolic) circle packing on a triangulated closed surface with prescribed intersection angles. His work generalizes Koebe's and Andreev's results of circle packing on a sphere [1, 2, 14]. Thurston conjectured that the discrete conformal mapping based on circle packing converges to the smooth Riemann mapping when the discrete tessellation becomes finer and finer. Thurston's conjecture has been proved by Rodin and Sullivan [21]. Chow and Luo established the intrinsic connection between circle packing and surface Ricci flow [6].

4.1 Discrete Surface

In practice, smooth surfaces are usually approximated by discrete surfaces, namely, triangular meshes. Figure 4.2 demonstrates such an example. The surface of the sculpture of Michelangelo's David is approximated by a triangular mesh. Each face of the mesh is a Euclidean triangle. With high sampling density, the discrete surface can approximate the smooth surface accurately.

The combinatorial structure of a discrete surface is represented by a simplicial complex. The fundamental concepts from smooth differential geometry, such as Riemannian metric, curvature and conformal structure, are generalized to the simplicial complex, respectively.

W. Zeng and X.D. Gu, *Ricci Flow for Shape Analysis and Surface Registration*,
SpringerBriefs in Mathematics, DOI 10.1007/978-1-4614-8781-4_4,
© Wei Zeng, Xianfeng David Gu 2013

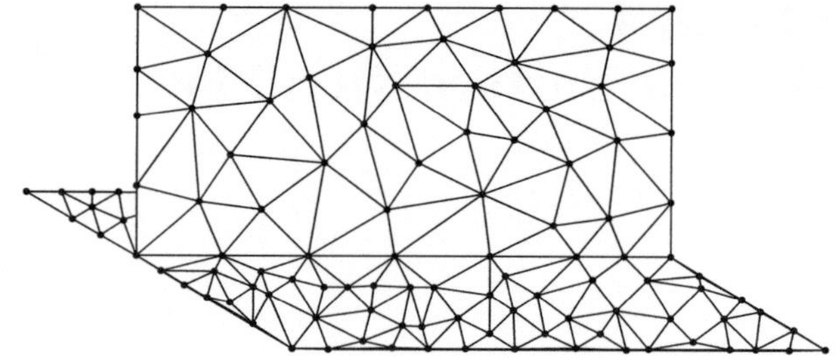

Fig. 4.1 A non-manifold simplicial complex

4.1.1 Simplicial Complex

Definition 4.1 (Simplex). Suppose $n + 1$ points $\{v_0, v_1, \ldots, v_n\}$ in the general positions in $\mathbb{R}^n$. The standard simplex $[v_0, v_1, \ldots, v_n]$ is the minimal convex set

$$\sigma = [v_0, v_1, \ldots, v_n] := \left\{ \sum_{i=0}^{n} \lambda_i v_i \mid \sum_{i=0}^{n} \lambda_i = 1, \lambda_i \geq 0 \right\}.$$

We call $v_0, v_1, \ldots, v_n$ the vertices of the simplex σ.

Definition 4.2 (Facet). Suppose σ is a simplex, and $\tau \subset \sigma$ is also a simplex. Then we say τ is a facet of σ.

The simplex has an orientation. Suppose $(i_0, i_1, \ldots, i_n)$ is a permutation of $(0, 1, \ldots, n)$. $[v_{i_0}, v_{i_1}, \ldots, v_{i_n}]$ has the same orientation as $[v_0, v_1, \ldots, v_n]$ if the permutation is the product of even number of swaps. Otherwise, they have opposite orientations if the permutation can be decomposed to odd number of swaps.

Definition 4.3 (Boundary Operator). The boundary of a simplex σ is

$$\partial \sigma = \sum_{i=0}^{n} (-1)^j [v_0, \ldots, v_{j-1}, v_{j+1}, \ldots, v_n].$$

The combinatorial structure of a discrete surface is usually represented as a simplicial complex.

Definition 4.4 (Simplicial Complex). A simplicial complex Σ is a union of simplicies, such that

1. If a simplex σ belongs to Σ, then all its facets also belong to Σ.
2. If two simplicies $\sigma_1, \sigma_2 \subset \Sigma$, $\sigma_1 \cap \sigma_2 \neq \emptyset$, then their intersection is a common facet.

A simplicial complex may not necessarily be a manifold. Figure 4.1 gives a counter example, where three half planes meet together at their boundaries.

Discrete surfaces are represented as two-dimensional simplicial complexes which are manifolds, as shown in Fig. 4.2.

Fig. 4.2 Smooth surfaces are approximated by discrete surfaces

Definition 4.5 (Triangular Mesh). Suppose Σ is a two-dimensional simplicial complex, furthermore it is also a manifold, namely, for each point p of Σ, there exists a neighborhood of p, $U(p)$, which is homeomorphic to the whole plane or the upper half plane. Then Σ is called a triangular mesh.

If $U(p)$ is homeomorphic to the whole plane, then p is called an interior point; if $U(p)$ is homeomorphic to the upper half plane, then p is called a boundary point.

The combinatorics of the simplicial complex fully determines the topology of the surface. The fundamental geometric quantities, such as Riemannian metric and curvatures, are represented as functions defined on the edges and vertices of the complex.

4.1.2   Discrete Riemannian Metric and Curvature

In the following discussion, we use $\Sigma = (V, E, F)$ to denote the mesh with vertex set V, edge set E, and face set F. We say that a discrete surface is with Euclidean (hyperbolic or spherical) background geometry if it is constructed by isometrically gluing triangles in $\mathbb{E}^2$ ($\mathbb{H}^2$ or $\mathbb{S}^2$).

We need to generalize the concept of Riemannian metric and curvatures to the discrete surfaces. Because the Riemannian metric on each triangle is totally determined by its three edge lengths, it is natural to define the discrete Riemannian metric as the edge lengths.

Definition 4.6 (Discrete Riemannian Metric). A discrete metric on a triangular mesh is a function defined on the edges, $l : E \rightarrow \mathbb{R}^+$, which satisfies the triangle inequality: on each face $[v_i, v_j, v_k]$,

$$l_{ij} + l_{jk} > l_{ki}, \; l_{jk} + l_{ki} > l_{ij}, \; l_{ki} + l_{ij} > l_{jk}.$$

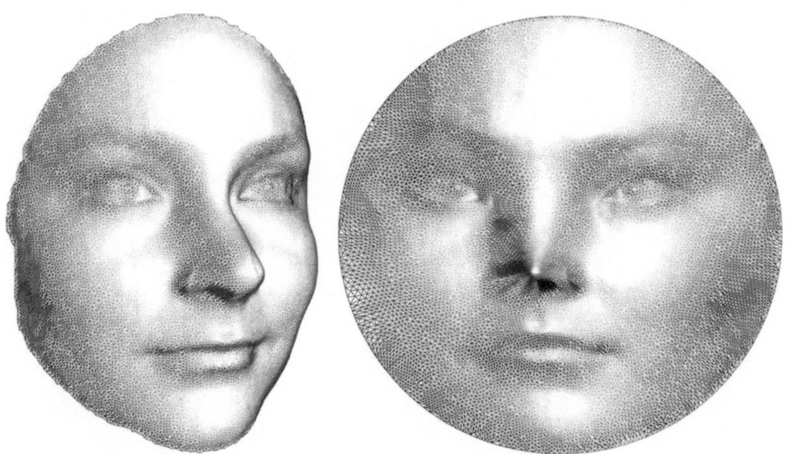

Fig. 4.3 A triangular mesh has different discrete Riemannian metrics

Fixing the combinatorial structure of a mesh, there are infinitely many discrete Riemannian metrics. Due to the fact that the convex combinations of discrete metrics are still discrete metrics, it is straightforward to show that the space of all discrete Riemannian metrics on a triangular mesh is convex. Figure 4.3 shows two different discrete metrics on the same triangular mesh, where the left one has 3D geometry while the right one is flat.

Definition 4.7 (Discrete Gauss Curvature).   The discrete Gauss curvature function on a mesh is defined on vertices, $K : V \to \mathbb{R}$,

$$K(v) = \begin{cases} 2\pi - \sum_i \alpha_i, & v \notin \partial M \\ \pi - \sum_i \alpha_i, & v \in \partial M \end{cases},$$

where α_i's are corner angles adjacent to the vertex v, and ∂M represents the boundary of the mesh.

Figure 4.4 gives the illustration for computing the discrete Gauss curvatures.

The discrete mean curvature function is defined on edges, which is the product of the edge length and the edge dihedral angle.

Definition 4.8 (Discrete Mean Curvature). Suppose $[v_0, v_1]$ is an edge with dihedral angle θ and length l. The mean curvature on the edge is given by

$$H([v_0, v_1]) = l\theta.$$

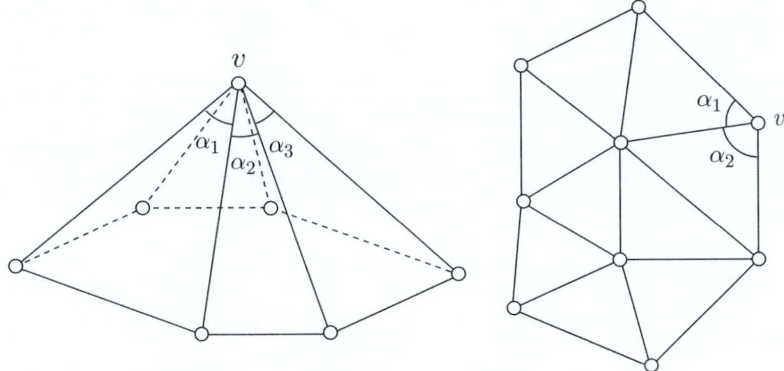

Fig. 4.4 Discrete curvatures of an interior vertex and a boundary vertex

4.1.3 Discrete Gauss–Bonnet Theorem

Although the Gauss curvature is determined by the Riemannian metric, Gauss–Bonnet theorem claims that the total curvature is a topological invariant. A discrete surface can be treated as a surface endowed with a flat Riemannian metric with cone singularities (at the vertices). The Gauss curvature is zero almost everywhere except at the vertices. Naturally, Gauss–Bonnet theorem should hold for discrete surfaces. In the following, we prove it in the setting of triangular mesh with Euclidean background metric.

Theorem 4.1 (Discrete Gauss–Bonnet Theorem). *Suppose Σ is a triangular mesh with Euclidean background metric. The total curvature is a topological invariant,*

$$\sum_{v \notin \partial \Sigma} K(v) + \sum_{v \in \partial \Sigma} K(v) = 2\pi \chi(\Sigma),$$

where χ is the characteristic Euler number and K is the Gauss curvature.

Proof. First, we assume the mesh Σ is closed without boundary components and take V, E, and F to be the numbers of vertices, edges, and faces, respectively. Then each face has 3 edges, and each edge is shared by 2 faces, therefore $3F = 2E$. The Euler characteristic number $\chi(\Sigma) = V + F - E = V - \frac{F}{2}$. Let v_i be a vertex, and the corner angle at v_i in face $[v_i, v_j, v_k]$ be α_i^{jk}, then the Gauss curvature of v_i is

$$K(v_i) = 2\pi - \sum_{jk} \alpha_i^{jk}.$$

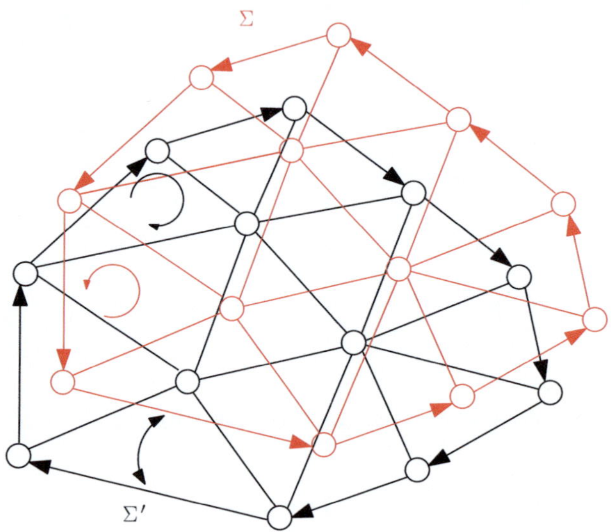

Fig. 4.5 Double covering

The total Gauss curvature is

$$\sum_i K(v_i) = \sum_i \left(2\pi - \sum_{jk} \alpha_i^{jk} \right) = 2\pi V - \sum_{ijk} (\alpha_i^{jk} + \alpha_j^{ki} + \alpha_k^{ij})$$

$$= 2\pi V - \pi F = 2\pi \left(V - \frac{F}{2} \right).$$

Second, we prove the theorem for meshes with boundary components by *double covering*. Suppose Σ has boundaries. As shown in Fig. 4.5, we construct a duplicate mesh Σ', such that each vertex $v_i \in \Sigma$ has a corresponding vertex $v_i' \in \Sigma'$, and each face $[v_i, v_j, v_k]$ has a corresponding face $[v_i', v_j', v_k']$. Then we reverse the orientation of all faces in Σ', namely, we change the order of the face vertices from $[v_i', v_j', v_k']$ to $[v_j', v_i', v_k']$. Then we identify each boundary vertex $v_i \in \partial\Sigma$ with the corresponding vertex $v_i' \in \partial\Sigma'$, which gives an equivalence relation $v_i \sim v_i'$. The doubled mesh $\bar{\Sigma}$ is given by the quotient space

$$\bar{\Sigma} := \{\Sigma \cup \Sigma'\}/\sim .$$

If $v_i \in \Sigma$ is an interior vertex in Σ, then it is included in $\bar{\Sigma}$, and the Gauss curvature of v_i in Σ is same as that in $\bar{\Sigma}$,

$$K_\Sigma(v_i) = K_{\bar{\Sigma}}(v_i), v_i \notin \partial\Sigma.$$

If $v_i \in \partial\Sigma$ is a boundary vertex in Σ, then by the definition (4.7) of discrete geodesic and Gauss curvatures, the Gauss curvature of v_i in $\bar{\Sigma}$ doubles its geodesic curvature in Σ,

$$2K_\Sigma(v_i) = K_{\bar{\Sigma}}(v_i).$$

Therefore the total curvature of $\bar{\Sigma}$ doubles the total curvature of Σ.

On the other hand, let E_0 represent the number of interior edges in Σ, E_1 the number of boundary edges, V_0 the number of interior vertices, and V_1 the number of boundary vertices. Then it is obvious that $V_1 = E_1$, and the Euler characteristic number of $\bar{\Sigma}$ is given by

$$\chi(\bar{\Sigma}) = 2V_0 + V_1 + 2F - (2E_0 + E_1) = 2(V_0 + V_1) + 2F - 2(E_0 + E_1) = 2\chi(\Sigma),$$

therefore,

$$2\sum_{v\in\Sigma} K_\Sigma(v) = \sum_{v\in\bar{\Sigma}} K_{\bar{\Sigma}}(v) = 2\pi\chi(\bar{\Sigma}) = 4\pi\chi(\Sigma).$$

The theorem holds for discrete surfaces with boundaries. $\square$

4.2   Euclidean Discrete Surface Ricci Flow

Discrete surface Ricci flow can be interpreted in a variational framework. Ricci flow is the negative gradient flow of a convex energy. The convexity of the energy induces the one-to-one correspondence between the curvature functions and the conformal metrics.

4.2.1   Thurston's Intuition

From smooth surface Ricci flow theory, we know surface Ricci flow conformally deforms the Riemannian metric, and conformal metric deformation preserves infinitesimal circles.

Figure 4.6 illustrates this unique characteristic of conformal deformation, where a human face surface is mapped onto the planar disk by a Riemann mapping. The regular circle packing on the plane (packed by circular disks with uniform sizes) is pulled back onto the face surface and forms a circle packing on the surface. The circular shapes are well preserved, but the sizes of circles are highly nonuniform, because locally, a conformal mapping is a similarity transformation and preserves local shapes, but distorts the areas.

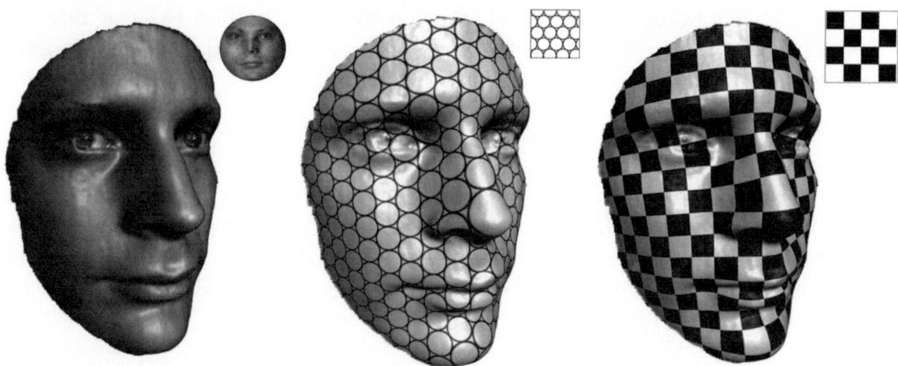

Fig. 4.6 Conformal mapping preserves infinitesimal circles

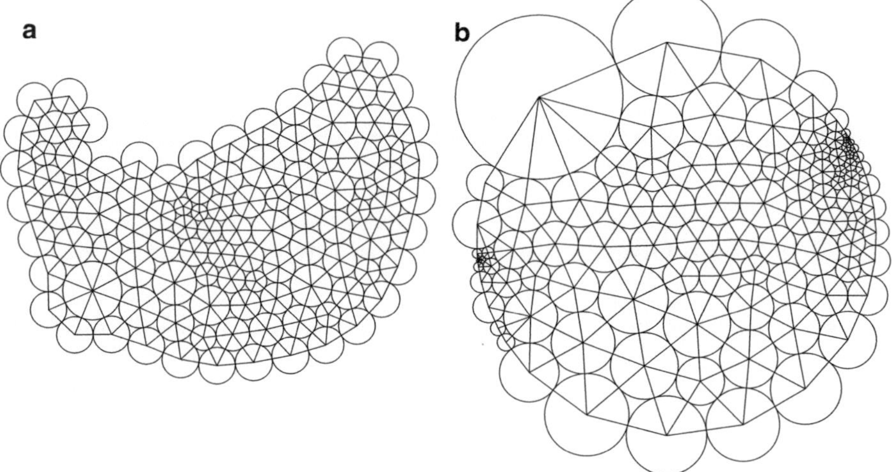

Fig. 4.7 Discrete Riemann mapping using circle packing. (**a**) Domain, (**b**) Range

Inspired by the nature of conformal metric deformation, Thurston proposed to use circle packing to approximate conformal mapping, which replaces infinitesimal circles to circles with finite sizes. Figure 4.7 demonstrates the principles. Suppose Ω is a planar simply connected domain. We would like to compute the Riemann mapping $\phi : \Omega \to \mathbb{D}$, which maps Ω to the unit disk. We triangulate the domain Ω and associate each vertex with a circle. For each edge $[v_i, v_j]$, the two circles centered at the two end vertices v_i and v_j are tangent to each other. We change the radii of all circles, preserving their tangential relations and the combinatorial structure of the triangulation. Then we deform the domain Ω to a convex polygon, which approximates the unit disk. The mapping constructed is a linear map on each triangle and also preserves the circle pattern. Thurston [25] conjectured that when

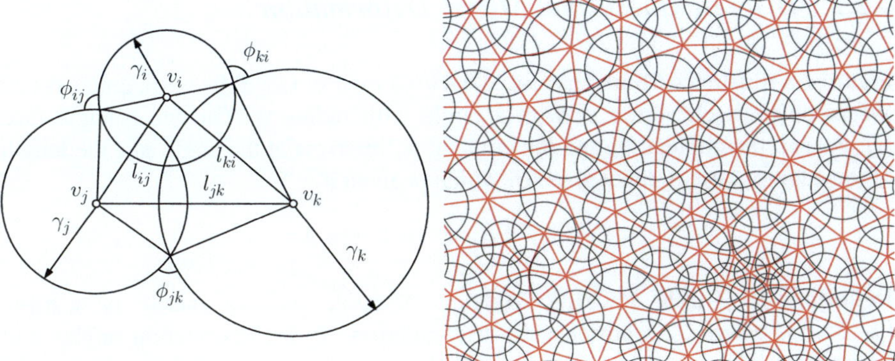

Fig. 4.8 Circle packing metric

the tessellation goes to infinitely finer, then the piecewise linear map converges to the real Riemann mapping. The conjecture was proved by Rodin and Sullivan [21].

Thurston's idea can be applied for surfaces as well. Given a surface, we compute a triangulation, then associate each vertex with a circle; the circles associated with the end vertices of an edge $[v_i, v_j]$ may intersect each other with a nonobtuse angle ϕ_{ij}, as shown in Fig. 4.8. In fact, each circle is a cone in $\mathbb{R}^3$, the circle center is the apex, and the radius is the lateral height. For simplicity, we still call these cones circles. The difference between 2π and the cone angle is called *discrete curvature* at the vertex. During the deformation, we only change the circle radii and keep the combinatorial structure of the triangulation and the intersection angles ϕ_{ij}'s. Then the discrete curvatures at the vertices (cone angles) will change accordingly. Intuitively, if we enlarge the radius at one vertex, then it becomes sharper and its curvature increases; similarly, if we reduce the radius, then it becomes flatter and its curvature decreases. Suppose we want to find a conformal mapping to "flatten" the face surface in Fig. 4.5 to the plane. Thurston's algorithm is as follows: At each step, we sort the vertices according to the absolute values of their curvatures, pick the one with the greatest absolute value. If its curvature is positive, then reduce its radius; if its curvature is negative, then increase its radius, until its curvature becomes 0. All the other radii are kept unchanged during the process. Then we repeat the procedure, eventually, all the interior vertices have zero curvatures. Namely, the surface is flat. Each triangle on the surface is mapped to a planar triangle. The piecewise linear map from the surface to the plane is an approximation of the Riemann mapping. If we subdivide the surface triangulation infinite times, then the discrete mappings converge to the real smooth Riemann mapping.

In the following, we discuss the discrete surface Ricci flow theory based on Thurston's circle packing. Thurston's circle packing has been generalized to many different configurations. The main theorems hold for all other variations as well. We will give a coherent framework to unify all the configurations in the later discussion.

4.2.2   Discrete Conformal Metric Deformation

Figure 4.8 illustrates Thurston's circle packing metric. Let Σ be a triangular mesh. We associate each vertex v_i with a circle with radius γ_i. On one triangle face $[v_i, v_j, v_k]$, the circle at v_i and the circle at v_j intersect at the angle ϕ_{ij}; the length of the edge $[v_i, v_j]$ is denoted as l_{ij}, which is given by

$$l_{ij}^2 = \gamma_i^2 + \gamma_j^2 + 2\gamma_i\gamma_j \cos\phi_{ij}.$$

Definition 4.9 (Circle Packing Metric). A circle packing metric is a triple (Σ, Φ, Γ), where Σ represents the triangulation, Φ the intersection angles and Γ the radii,

$$\Phi = \{\phi_{ij} | \forall e_{ij}\}, \quad \Gamma = \{\gamma_i | \forall v_i\}.$$

In practice, it is more convenient to use the logarithm of the circle radius, which is the discrete analogy to the concept of conformal factor in the smooth case.

Definition 4.10 (Discrete Conformal Factor). Discrete conformal factor on a mesh Σ is a function defined on each vertex $\mathbf{u} : V \to \mathbb{R}$,

$$u_i = \log\gamma_i.$$

In the smooth case, changing a Riemannian metric by a scalar function, $\mathbf{g} \to e^{2u}\mathbf{g}$, is called a conformal metric deformation. The discrete analogy to this is as follows.

Definition 4.11 (Discrete Conformal Equivalence). Two circle packing metrics $(\Sigma_k, \Phi_k, \Gamma_k)$, $k = 1, 2$, are conformally equivalent if $\Sigma_1 = \Sigma_2$ and $\Phi_1 = \Phi_2$.

We can also say that the circle packing metric $(\Sigma_1, \Phi_1, \Gamma_1)$ is deformed to $(\Sigma_2, \Phi_2, \Gamma_2)$ conformally. Namely, if we only change the circle radii and keep the combinatorial structure and the intersection angles, then the discrete metric deformation is conformal.

4.2.3   Euclidean Derivative Cosine Law

On a smooth surface, the Riemannian metric determines its Gauss curvature. This also holds for discrete surfaces. The edge lengths determine the corner angles by cosine law. During the Ricci flow, the metric deforms smoothly, therefore we need to study the derivative cosine law.

Suppose $[v_i, v_j, v_k]$ is a Euclidean triangle, as shown in Fig. 4.9. We treat each corner angle $\theta_i, \theta_j, \theta_k$ as the function of edge lengths l_i, l_j, l_k, respectively. Then the derivative of $\theta_i(l_i, l_j, l_k)$ with respect to the edge lengths l_i, l_j, l_k are given by the following lemma.

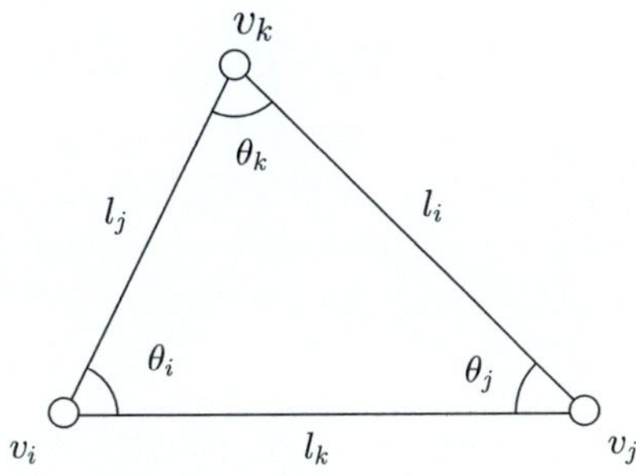

Fig. 4.9 A Euclidean triangle

Lemma 4.1 (Derivative Cosine Law).

$$\frac{\partial \theta_i}{\partial l_i} = \frac{l_i}{A}, \quad \frac{\partial \theta_i}{\partial l_j} = -\frac{\partial \theta_i}{\partial l_i} \cos \theta_k,$$

where $A = l_j l_k \sin \theta_i$ is the double of the triangle area.

Proof. The proof is by direct computation. From the Euclidean cosine law,

$$2 l_j l_k \cos \theta_i = l_j^2 + l_k^2 - l_i^2.$$

On both sides, taking the derivative with respect to l_i, we obtain $\frac{\partial \theta_i}{\partial l_i} = \frac{l_i}{A}$; taking the derivative with respect to l_j, we get

$$\frac{\partial \theta_i}{\partial l_j} = -\frac{l_i \cos \theta_k}{A} = -\frac{\partial \theta_i}{\partial l_i} \cos \theta_k. \qquad \qquad \square$$

We rewrite the lemma in the matrix format,

$$\begin{pmatrix} d\theta_i \\ d\theta_j \\ d\theta_k \end{pmatrix} = \frac{-1}{A} \begin{pmatrix} l_i & 0 & 0 \\ 0 & l_j & 0 \\ 0 & 0 & l_k \end{pmatrix} \begin{pmatrix} -1 & \cos \theta_k & \cos \theta_j \\ \cos \theta_k & -1 & \cos \theta_i \\ \cos \theta_j & \cos \theta_i & -1 \end{pmatrix} \begin{pmatrix} dl_i \\ dl_j \\ dl_k \end{pmatrix}. \qquad (4.1)$$

Suppose the triangle is with a circle packing, then we define the power center as follows.

Definition 4.12 (Power Center). Given a triangle $[v_i, v_j, v_k]$ with a circle packing, the unique circle orthogonal to three circles at the vertices is called the *power circle*. The center of the power circle is called the *power center*.

As shown in Fig. 4.10, two intersecting circles have a common chord. Three common chords intersect at the power center o. Through the power center, we draw three lines perpendicular to three edges, h_i, h_j, h_k, intersecting with the edges at

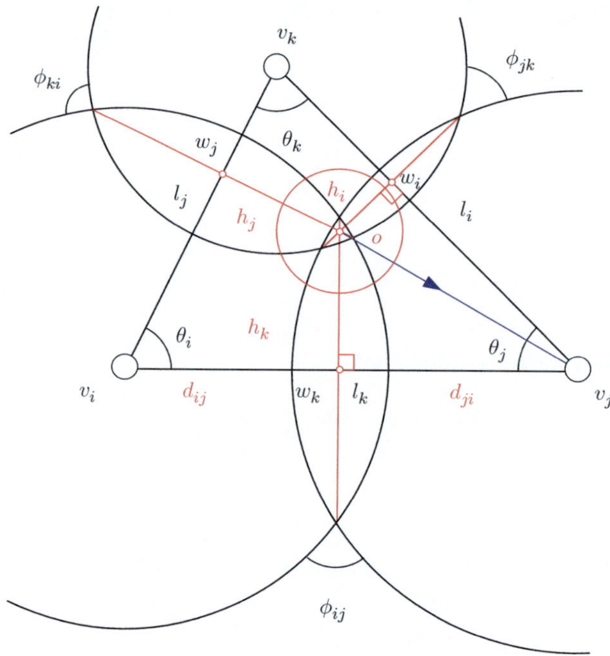

Fig. 4.10 Derivative cosine law

the pedal points w_i, w_j, w_k, respectively. Each pedal point divides the edge to two segments. The distance from v_i to w_k is d_{ij} and the distance from v_j to w_k is d_{ji}.

The edge lengths are the functions of the discrete conformal factors (the logarithms of circle radii). The derivatives can be obtained by direct computation.

Lemma 4.2. *In triangle $[v_i, v_j, v_k]$ with a circle packing,*

$$\frac{\partial l_k}{\partial u_j} = d_{ji}.$$

Proof. See Fig. 4.11**a**. According to the Euclidean cosine law,

$$l_k^2 = r_i^2 + r_j^2 + 2\cos\phi_{ij}\, r_i r_j,$$

taking the derivative with respect to l_k on both sides, we get

$$\frac{\partial l_k}{\partial r_j} = \frac{r_j + r_i \cos\phi_{ij}}{l_k},$$

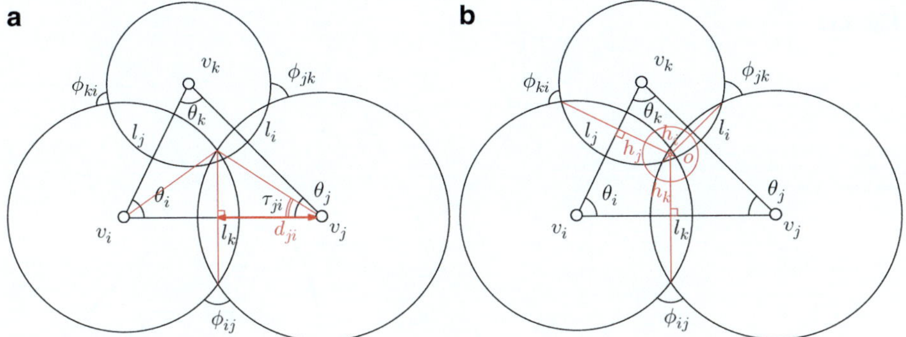

Fig. 4.11 Derivative relation between edge length (or corner angle) and conformal factor

therefore

$$\frac{\partial l_k}{\partial u_j} = r_j \frac{\partial l_k}{\partial r_j} = \frac{2r_j^2 + 2r_i r_j \cos\phi_{ij}}{2l_k} = \frac{l_k^2 + r_j^2 - r_i^2}{2l_k}$$

$$= 2\frac{l_k r_j \cos\tau_{ji}}{2l_k} = r_j \cos\tau_{ji} = d_{ji}.$$

$\square$

Namely,

$$\begin{pmatrix} dl_i \\ dl_j \\ dl_k \end{pmatrix} = \begin{pmatrix} \frac{1}{2l_i} & 0 & 0 \\ 0 & \frac{1}{2l_j} & 0 \\ 0 & 0 & \frac{1}{2l_k} \end{pmatrix} \begin{pmatrix} 0 & l_i^2 + r_j^2 - r_k^2 & l_i^2 + r_k^2 - r_j^2 \\ l_j^2 + r_i^2 - r_k^2 & 0 & l_j^2 + r_k^2 - r_i^2 \\ l_k^2 + r_i^2 - r_j^2 & l_k^2 + r_j^2 - r_i^2 & 0 \end{pmatrix} \begin{pmatrix} du_i \\ du_j \\ du_k \end{pmatrix}.$$

$$(4.2)$$

Furthermore,

$$\begin{pmatrix} dl_i \\ dl_j \\ dl_k \end{pmatrix} = \begin{pmatrix} 0 & d_{jk} & d_{kj} \\ d_{ik} & 0 & d_{ki} \\ d_{ij} & d_{ji} & 0 \end{pmatrix} \begin{pmatrix} du_i \\ du_j \\ du_k \end{pmatrix}.$$

$$(4.3)$$

Now we can show a property for d_{ij}.

Lemma 4.3. *For the triangle $[v_i, v_j, v_k]$ with a circle packing,*

$$d_{ij}^2 + d_{jk}^2 + d_{ki}^2 = d_{ji}^2 + d_{kj}^2 + d_{ik}^2.$$

Proof. By the Pythagorean theorem,

$$d_{ij}^2 + h_k^2 = d_{ik}^2 + h_j^2, \quad d_{jk}^2 + h_i^2 = d_{ji}^2 + h_k^2, \quad d_{ki}^2 + h_j^2 = d_{kj}^2 + h_i^2.$$

Adding both sides, we complete the proof.

$\square$

Fig. 4.12 $\dfrac{\partial \theta_i}{\partial u_j} = \dfrac{h_k}{l_k}$

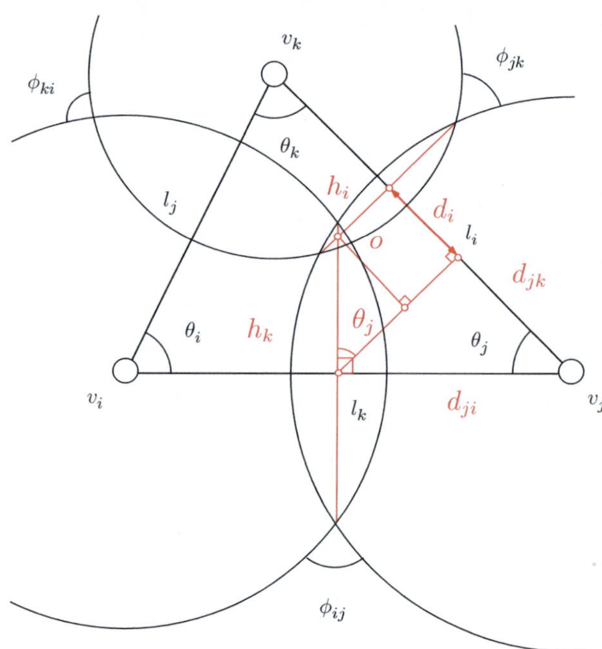

The following symmetry lemma is crucial to the whole discrete Ricci flow theory.

Lemma 4.4 (Symmetry). *In a triangle $[v_i, v_j, v_k]$ with a circle packing,*

$$\frac{\partial \theta_i}{\partial u_j} = \frac{\partial \theta_j}{\partial u_i} = \frac{h_k}{l_k}, \quad \frac{\partial \theta_j}{\partial u_k} = \frac{\partial \theta_k}{\partial u_j} = \frac{h_i}{l_i}, \quad \frac{\partial \theta_k}{\partial u_i} = \frac{\partial \theta_i}{\partial u_k} = \frac{h_j}{l_j}, \tag{4.4}$$

and

$$\frac{\partial \theta_i}{\partial u_i} = -\frac{\partial \theta_i}{\partial u_j} - \frac{\partial \theta_i}{\partial u_k}, \quad \frac{\partial \theta_j}{\partial u_j} = -\frac{\partial \theta_j}{\partial u_k} - \frac{\partial \theta_j}{\partial u_i}, \quad \frac{\partial \theta_k}{\partial u_k} = -\frac{\partial \theta_k}{\partial u_i} - \frac{\partial \theta_k}{\partial u_j}. \tag{4.5}$$

Proof. See Fig. 4.11**b**. First we prove the equations in (4.4). From Lemma 4.1,

$$\frac{\partial \theta_i}{\partial u_j} = \frac{\partial \theta_i}{\partial l_i} \frac{\partial l_i}{\partial u_j} + \frac{\partial \theta_i}{\partial l_k} \frac{\partial l_k}{\partial u_j} = \frac{\partial \theta_i}{\partial l_i} \left(\frac{\partial l_i}{\partial u_j} - \frac{\partial l_k}{\partial u_j} \cos \theta_j \right).$$

From Lemma 4.2, the right-hand side of the above equation equals

$$\frac{l_i}{A}(d_{jk} - d_{ji} \cos \theta_j) = \frac{d_i l_i}{l_i l_k \sin \theta_j} = \frac{h_k \sin \theta_j}{l_k \sin \theta_j} = \frac{h_k}{l_k},$$

where $d_i = d_{jk} - d_{ji} \cos \theta_j$, as shown in Fig. 4.12.

Then we prove the equations in (4.5). Because $\theta_i + \theta_j + \theta_k = \pi$,

$$\frac{\partial \theta_i}{\partial u_i} = -\frac{\partial \theta_j}{\partial u_i} - \frac{\partial \theta_k}{\partial u_i} = -\frac{\partial \theta_i}{\partial u_j} - \frac{\partial \theta_i}{\partial u_k}.$$

$\square$

We formulate the lemma in the matrix format by combining the matrix formulae (4.1) and (4.3),

$$\begin{pmatrix} d\theta_i \\ d\theta_j \\ d\theta_k \end{pmatrix} = - \begin{pmatrix} \frac{h_k}{l_k} + \frac{h_j}{l_j} & -\frac{h_k}{l_k} & -\frac{h_j}{l_j} \\ -\frac{h_k}{l_k} & \frac{h_k}{l_k} + \frac{h_i}{l_i} & -\frac{h_i}{l_i} \\ -\frac{h_j}{l_j} & -\frac{h_i}{l_i} & \frac{h_j}{l_j} + \frac{h_i}{l_i} \end{pmatrix} \begin{pmatrix} du_i \\ du_j \\ du_k \end{pmatrix}. \tag{4.6}$$

Now let's consider the admissible space of (u_i, u_j, u_k).

Lemma 4.5 (Admissible Space). *For any three nonobtuse angles* $\phi_{ij}, \phi_{jk}, \phi_{ki} \in [0, \frac{\pi}{2}]$ *and any three positive numbers* r_i, r_j, r_k, *there is a configuration of three circles in Euclidean geometry, unique up to isometry, having radii* (r_i, r_j, r_k) *and meeting in angles* $(\phi_{ij}, \phi_{jk}, \phi_{ki})$.

Proof. It is sufficient and necessary to show that the triangle inequality holds.

$$\max\{r_i^2, r_j^2\} < r_i^2 + r_j^2 + 2r_i r_j \cos\phi_{ij} \le (r_i + r_j)^2,$$

$$\max\{r_i, r_j\} < l_k \le r_i + r_j,$$

so

$$l_k \le r_i + r_j < l_i + l_j.$$

$\square$

Therefore, if the angles $\phi_{ij}, \phi_{jk}, \phi_{ki}$ are nonobtuse, then the admissible space for (u_i, u_j, u_k) is $\mathbb{R}^3$.

Lemma 4.6. *The differential form*

$$\omega = \theta_i du_i + \theta_j du_j + \theta_k du_k$$

is an exact 1-form.

Proof. Because $\frac{\partial \theta_i}{\partial u_j} = \frac{\partial \theta_j}{\partial u_i}$,

$$d\omega = \left(\frac{\partial \theta_i}{\partial u_j} - \frac{\partial \theta_j}{\partial u_i}\right) du_j \wedge du_i + \left(\frac{\partial \theta_j}{\partial u_k} - \frac{\partial \theta_k}{\partial u_j}\right) du_k \wedge du_j$$

$$+ \left(\frac{\partial \theta_k}{\partial u_i} - \frac{\partial \theta_i}{\partial u_k}\right) du_i \wedge du_k = 0.$$

Therefore ω is a closed 1-form. Furthermore, the admissible space for (u_i, u_j, u_k) is $\mathbb{R}^3$, which is simply connected, so ω is exact. $\square$

Because ω is an exact 1-form, it is the gradient of a function, which is the discrete Ricci energy defined on all the circle packing metrics on the single triangle with intersection angles $(\phi_{ij}, \phi_{jk}, \phi_{ki})$.

Definition 4.13 (Discrete Ricci Energy). The discrete Ricci energy is given by

$$E(u_i, u_j, u_k) = \int_{(0,0,0)}^{(u_i, u_j, u_k)} \omega.$$

Theorem 4.2 (Concavity of Discrete Ricci Energy). *The Ricci energy $E(u_i, u_j, u_k)$ is strictly concave on the subspace $u_i + u_j + u_k = 0$.*

Proof. From Lemma 4.6, the gradient $\nabla E = (\theta_i, \theta_j, \theta_k)$. From Lemma 4.4, the Hessian matrix is

$$H = \begin{pmatrix} \frac{\partial \theta_i}{\partial u_i} & \frac{\partial \theta_i}{\partial u_j} & \frac{\partial \theta_i}{\partial u_k} \\ \frac{\partial \theta_j}{\partial u_i} & \frac{\partial \theta_j}{\partial u_j} & \frac{\partial \theta_j}{\partial u_k} \\ \frac{\partial \theta_k}{\partial u_i} & \frac{\partial \theta_k}{\partial u_j} & \frac{\partial \theta_k}{\partial u_k} \end{pmatrix} = - \begin{pmatrix} \frac{h_k}{l_k} + \frac{h_j}{l_j} & -\frac{h_k}{l_k} & -\frac{h_j}{l_j} \\ -\frac{h_k}{l_k} & \frac{h_k}{l_k} + \frac{h_i}{l_i} & -\frac{h_i}{l_i} \\ -\frac{h_j}{l_j} & -\frac{h_i}{l_i} & \frac{h_j}{l_j} + \frac{h_i}{l_i} \end{pmatrix}.$$

The negative of the Hessian matrix $-H$ is diagonal dominant, with null space $(1, 1, 1)$. On the subspace $u_i + u_j + u_k = 0$, H is strictly negative definite. Therefore the discrete Ricci energy $E(u_i, u_j, u_k)$ is strictly concave. $\qquad\square$

4.2.4　Discrete Ricci Energy

Now we generalize the discrete Ricci energy to the whole triangular mesh.

Definition 4.14 (Edge Weight). Suppose $[v_i, v_j]$ is an interior edge of a triangular mesh Σ, adjacent to two faces $[v_i, v_j, v_k]$ and $[v_j, v_i, v_l]$. Suppose the mesh is with a circle packing metric. Each face has a power center, o_k for $[v_i, v_j, v_k]$ and o_l for $[v_j, v_i, v_l]$. The distances from o_k, o_l to $[v_i, v_j]$ are h_k and h_l, respectively. The length of $[v_i, v_j]$ is l_k. After the two faces are flattened, the edge weight for $[v_i, v_j]$ is given by

$$w_{ij} = \frac{|o_k - o_l|}{|v_i - v_j|} = \frac{h_k + h_l}{l_k}.$$

If $[v_i, v_j]$ is a boundary edge, draw a line from o_k perpendicular to the edge, and the pedal point is w_k, then the edge weight is given by

$$w_{ij} = \frac{|o_k - w_k|}{|v_i - v_j|} = \frac{h_k}{l_k}.$$

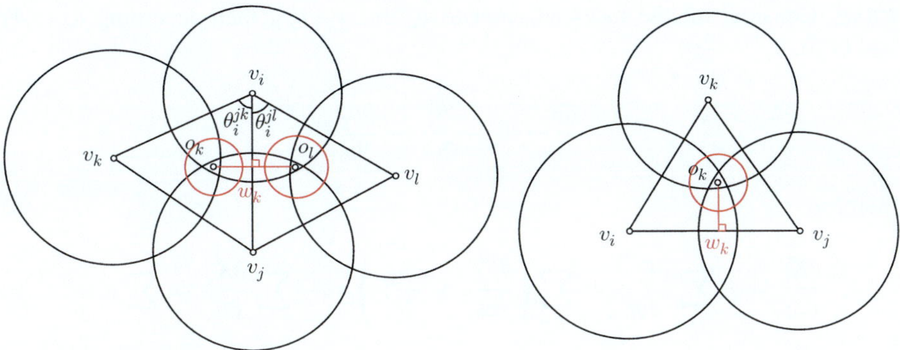

Fig. 4.13 Edge weight w_{ij}

Lemma 4.7. *Suppose $[v_i, v_j]$ is an edge on the mesh Σ with a circle packing metric. Then*

$$\frac{\partial K_i}{\partial u_j} = \frac{\partial K_j}{\partial u_i} = -w_{ij}. \tag{4.7}$$

Proof. See Fig. 4.13. Suppose $[v_i, v_j]$ is an interior edge. According to (4.4),

$$\frac{\partial \theta_i^{jk}}{\partial u_j} = \frac{|o_k - w_k|}{|v_i - v_j|}, \frac{\partial \theta_i^{jl}}{\partial u_j} = \frac{|o_l - w_k|}{|v_i - v_j|},$$

we have

$$\frac{\partial K_i}{\partial u_j} = -\frac{\partial \theta_i^{jk}}{\partial u_j} - \frac{\partial \theta_i^{jl}}{\partial u_j} = -\frac{|o_k - o_l|}{|v_i - v_j|} = -w_{ij}.$$

Similarly, if $[v_i, v_j]$ is a boundary edge, then

$$\frac{\partial K_i}{\partial u_j} = -\frac{\partial \theta_i^{jk}}{\partial u_j} = -\frac{|o_k - w_k|}{|v_i - v_j|} = -w_{ij}.$$

Therefore, we get

$$\frac{\partial K_i}{\partial u_j} = \frac{\partial K_j}{\partial u_i}.$$

$\square$

Lemma 4.8. *Suppose v_i is a vertex of a mesh Σ with a circle packing metric. Then*

$$\frac{\partial K_i}{\partial u_i} = -\sum_{j \neq i} \frac{\partial K_i}{\partial u_j} = \sum_{j \neq i} w_{ij}. \tag{4.8}$$

Proof. Consider all the faces adjacent to v_i, $[v_i, v_j, v_k]$, then according to (4.4) and (4.5),

$$\frac{\partial \theta_i^{jk}}{\partial u_i} = -\frac{\partial \theta_j^{ki}}{\partial u_i} - \frac{\partial \theta_k^{ij}}{\partial u_i},$$

therefore

$$\frac{\partial K_i}{\partial u_i} = -\sum_{jk} \frac{\partial \theta_i^{jk}}{\partial u_i} = \sum_{jk} \left(\frac{\partial \theta_j^{ki}}{\partial u_i} + \frac{\partial \theta_k^{ij}}{\partial u_i} \right) = -\sum_j \frac{\partial K_j}{\partial u_i} = \sum_j w_{ij}.$$

$\square$

Definition 4.15 (Admissible Metric Space). Suppose Σ is a triangular mesh. Fix the circle intersection angles $\Phi = \{\phi_{ij}\}$. Let the vector of logarithms of circle radii be $\mathbf{u} = (u_1, u_2, \ldots, u_n)^T$. Then if the triangle inequality holds on all faces of Σ, namely, (Σ, Φ, Γ) is a circle packing metric, then we say $\mathbf{u}$ is admissible. The space of all admissible $\mathbf{u}$ vectors is called the admissible metric space for (Σ, Φ), denoted as $U(\Sigma, \Phi)$.

Definition 4.16 (Curvature Map). Given an admissible metric $\mathbf{u} \in U(\Sigma, \Phi)$, it induces the Gauss curvature on the mesh. The Gauss curvatures on all vertices are represented as a vector $\mathbf{k} = (K_1, K_2, \ldots, K_n) \in \mathbb{R}^n$. The mapping from metric $\mathbf{u}$ to the curvature vector $\mathbf{k}$ is called the curvature map, denoted as $\kappa : U(\Sigma, \Phi) \to \mathbb{R}^n$.

The image of the curvature map is called the admissible curvature space.

Definition 4.17 (Admissible Curvature Space). Suppose Σ is a triangular mesh. Fix the circle intersection angles $\Phi = \{\phi_{ij}\}$. Then the space of all possible Gauss curvatures $\mathbf{k} = (K_1, K_2, \ldots, K_n)$, which is induced by an admissible metric $\mathbf{u} \in U(\Sigma, \Phi)$, is called the admissible curvature space, denoted as $K(\Sigma, \Phi)$,

$$K(\Sigma, \phi) := \kappa(U(\Sigma, \Phi)).$$

Due to the discrete Gauss–Bonnet Theorem 4.1, the total curvature is determined by the topology of Σ. Therefore

$$K(\Sigma, \phi) \subset \left\{ \sum_{i=1}^{n} K_i = 2\pi \chi(\Sigma) \right\}.$$

Lemma 4.9 (Admissible Metric Space). *Given a triangular mesh Σ, fix circle intersection angles Φ, such that all ϕ_{ij}'s are nonobtuse angles, $\phi_{ij} \in [0, \frac{\pi}{2}]$, then the admissible metric space for (Σ, Φ) is $\mathbb{R}^n$,*

$$U(\Sigma, \Phi) = \mathbb{R}^n.$$

Proof. According to Lemma 4.5, for each face $[v_i, v_j, v_k]$, the admissible metric space $U([v_i, v_j, v_k], [\phi_{ij}, \phi_{jk}, \phi_{ki}])$ is $\mathbb{R}^n$. The admissible metric space of the whole mesh Σ is the intersection of the admissible metric spaces of all faces,

$$U(\Sigma, \Phi) = \bigcap_{[v_i, v_j, v_k] \in \Sigma} U([v_i, v_j, v_k], [\phi_{ij}, \phi_{jk}, \phi_{ki}]) = \mathbb{R}^n.$$

$\square$

Corollary 4.1. *The differential form*

$$\omega = \sum_{i=1}^{n} K_i \, d u_i$$

is an exact 1-form.

Proof. Lemma 4.7 shows ω is closed. Because the admissible metric space is simply connected, ω is exact. $\square$

So we can define the Ricci energy, using ω as the gradient.

Definition 4.18 (Discrete Ricci Energy). The discrete Ricci energy is defined on the admissible metric space for (Σ, Φ),

$$E(u_1, u_2, \ldots, u_n) = \int_{(0,0,\ldots,0)}^{(u_1, u_2, \ldots, u_n)} \omega.$$

Theorem 4.3 (Convexity of Ricci Energy). *The discrete Ricci energy is strictly convex on the space,*

$$\sum_{i=1}^{n} u_i = 0.$$

Proof. Here we give two proofs. The first proof is based on the concavity of the discrete Ricci energy on each triangle. Suppose there are V_0 interior vertices and V_1 boundary vertices. Let $E(\Sigma)$ represent the Ricci energy of the whole mesh and $E([v_i, v_j, v_k])$ the energy on the face. Then

$$E(\Sigma) = \sum_{v_i \notin \partial \Sigma} 2\pi u_i + \sum_{v_j \in \partial \Sigma} \pi u_j - \sum_{[v_i, v_j, v_k] \in \Sigma} E([v_i, v_j, v_k]).$$

The linear terms won't affect the convexity. For all the faces, according to Theorem 4.2, $E([v_i, v_j, v_k])$ is strictly concave on the space $u_i + u_j + u_k = 0$. The null space of the Hessian of $E([v_i, v_j, v_k])$ is $u_i = u_j = u_k$. Therefore, the null space of the Hessian of $E(\Sigma)$ is the intersection of all the face null spaces, namely, the one dimensional space spanned by $(1, 1, 1, \ldots, 1)^T$. In the complement space $\sum_i u_i = 0$, $E(\Sigma)$ is the negative sum of concave functions $E([v_i, v_j, v_k])$, so it is strictly convex.

The second proof is to directly compute the Hessian matrix of the energy. The gradient of $E(\Sigma)$ is

$$\nabla E(\Sigma) = (K_1, K_2, \ldots, K_n)^T.$$

Therefore, the element of the Hessian matrix

$$\frac{\partial^2 E(\Sigma)}{\partial u_i \partial u_j} = \frac{\partial K_i}{\partial u_j}.$$

According to Lemmas 4.7 and 4.8, if v_i is adjacent to v_j, then $\frac{\partial K_i}{\partial u_j} = -w_{ij}$, otherwise $\frac{\partial K_i}{\partial u_j} = 0$. The diagonal elements $\frac{\partial K_i}{\partial u_i} = \sum_{j \neq i} w_{ij}$. Therefore, the Hessian matrix is diagonal dominant, with one dimensional null space, spanned by $(1, 1, \ldots, 1)^T$. In the space $\sum_i u_i = 0$, the energy is strictly convex. $\square$

Note that the differential of the curvatures and the logarithm of radius satisfy the Laplace equation

$$dK_i = \sum_j w_{ij}(du_i - du_j).$$

The admissible curvature space is also convex, which can be described by the following lemma.

Lemma 4.10 (Admissible Curvature Space). *Suppose (Σ, Φ) is a triangular mesh with a circle packing, and all the circle intersection angles are nonobtuse, $\forall \phi_{ij} \in \Phi$, $\phi_{ij} \in [0, \frac{\pi}{2}]$. Let I be a proper subset of vertices V. F_I is the collection of faces, whose vertices are in I. The link of I, denoted as $Lk(I)$, is the set of pairs (e, v), where e is an edge whose vertices are not in I, and vertex $v \in I$, so that (e, v) form a triangle. Then for any admissible curvature, the following inequality holds,*

$$\sum_{v_i \in I} K_i > - \sum_{(e,v) \in Lk(I)} (\pi - \phi(e)) + 2\pi \chi(F_I). \tag{4.9}$$

Given a curvature vector $\mathbf{k} = (K_1, K_2, \ldots, K_n)$, *if for any proper subset I the inequality (4.18) holds, then the curvature is admissible.*

The details of the proof can be found in [6].

4.2.5 Global Rigidity

In the smooth case, the Gauss curvature is determined by the Riemannian metric. The inverse is not true. Different Riemannian metrics may induce the same Gauss

Fig. 4.14 Legendre
transformation

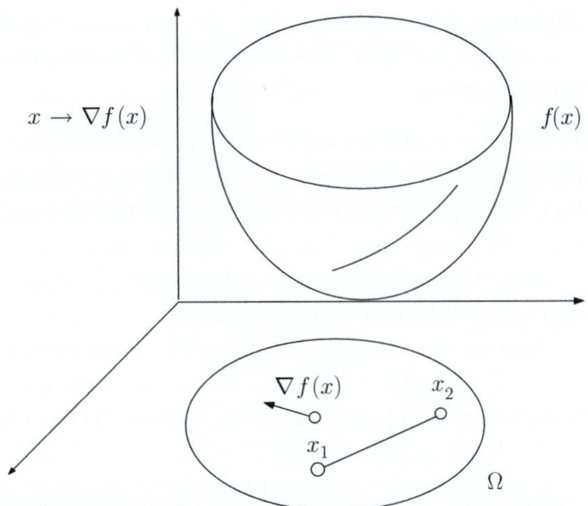

curvature function, but if the Riemannian metrics are restricted within a conformal
class, then the Gauss curvature can essentially determine the metric. Namely, the
solution to the Yamabe problem,

$$\begin{cases} \bar{K} = e^{-2\lambda}(K - \Delta_{\mathbf{g}}\lambda) \\ \bar{k}_g = e^{-\lambda}(k_g - \partial_{\mathbf{n},\mathbf{g}}\lambda) \end{cases},$$

is essentially unique. In the discrete case, the discrete metrics determine the discrete
curvatures. Inversely, the discrete curvatures determine the discrete conformal
Riemannian metrics. This uniqueness is called the *rigidity*. In the following, the
rigidity will be proved based on the *Legendre transformation*.

Legendre Transformation

Suppose $\Omega \subset \mathbb{R}^n$ is a convex domain. A C^2 function $f : \Omega \rightarrow \mathbb{R}$ is a convex
function, such that its Hessian matrix

$$\left(\frac{\partial^2 f}{\partial x_i \partial x_j} \right) > 0$$

is positive definite.

Definition 4.19 (Legendre Transformation). The gradient map

$$\phi : \mathbf{x} \rightarrow \nabla f(\mathbf{x}) = \left(\frac{\partial f}{\partial x_1}, \frac{\partial f}{\partial x_2}, \dots, \frac{\partial f}{\partial x_n} \right)^T$$

is called a Legendre transformation. See Fig. 4.14.

Theorem 4.4. *Suppose $\Omega \subset \mathbb{R}^n$ is a convex domain. A C^2 function $f : \Omega \to \mathbb{R}$ is a convex function. Then the Legendre transformation $\phi : \Omega \to \phi(\Omega)$ is a global diffeomorphism.*

Proof. The Jacobian matrix of ϕ is the Hessian of f, which is positive definite and invertible. Therefore ϕ is a local diffeomorphism, namely, for each point $\mathbf{p} \in \Omega$, there exists a neighborhood of $\mathbf{p}$, such that the restriction of ϕ on the neighborhood is a diffeomorphism. But globally, the mapping may not be one-to-one.

Now we need to show $\phi : \Omega \to \phi(\Omega)$ is a global diffeomorphism. It is necessary and sufficient to prove the injectivity.

Assume the map is not injective, there exist $\mathbf{p}_1, \mathbf{p}_2 \subset \Omega$, $\mathbf{p}_1 \neq \mathbf{p}_2$, such that $\nabla f(\mathbf{p}_1) = \nabla f(\mathbf{p}_2)$. Because Ω is convex, the line segment $(1-t)\mathbf{p}_1 + t\mathbf{p}_2, t \in [0, 1]$ is contained in Ω. The function $g(t) = f((1 - t)\mathbf{p}_1 + t\mathbf{p}_2)$ is convex, because

$$\frac{d^2 g(t)}{dt^2} = (\mathbf{p}_2 - \mathbf{p}_1)^T H (\mathbf{p}_2 - \mathbf{p}_1) > 0,$$

where H is the Hessian matrix of f. Therefore $\frac{dg(t)}{dt}$ is monotonous, but

$$g'(0) = \langle \nabla f(\mathbf{p}_1), \mathbf{p}_2 - \mathbf{p}_1 \rangle = \langle \nabla f(\mathbf{p}_2), \mathbf{p}_2 - \mathbf{p}_1 \rangle = g'(1)$$

is a contradiction. Therefore, the mapping is injective and is a global diffeomorphism. This completes the proof. $\qquad\square$

Now we come to the global rigidity for discrete Ricci flow.

Theorem 4.5 (Global Rigidity). *Suppose Σ is a triangular mesh, with circle packing metrics (Σ, Φ, Γ), where all the circle intersection angles Φ are fixed and are nonobtuse. Then the curvature map*

$$\kappa : U(\Sigma, \Phi) \bigcap \left\{ \sum_i u_i = 0 \right\} \to K(\Sigma, \Phi)$$

is a global diffeomorphism.

Proof. Consider the discrete Ricci energy $E : U(\Sigma, \Phi) \to \mathbb{R}$, which is with C^2 continuity. By Lemma 4.9, the domain $U(\Sigma, \Phi)$ is convex, so is the intersection between $U(\Sigma, \Phi)$ and the hyperplane $\sum_i u_i = 0$. By Theorem 4.3, on $U(\Sigma, \Phi) \bigcap \{\sum_i u_i = 0\}$ the energy is convex. The Legendre transformation induced by the energy is the curvature map

$$\kappa : (u_1, u_2, \ldots, u_n) \to \nabla E = (K_1, K_2, \ldots, K_n).$$

By the discrete Gauss–Bonnet Theorem 4.1, the total curvature $\sum_i K_i = 2\pi \chi(\Sigma)$. By Theorem 4.4, the curvature map is a global diffeomorphism. $\qquad\square$

4.2.6 Convergence Analysis

The discrete surface Ricci flow is given by

$$\frac{dr_i}{dt} = -K_i r_i,$$

or equivalently, $u_i = \log r_i$,

$$\frac{du_i}{dt} = -K_i.$$

In practice, it is useful to consider the normalized Ricci flow.

Definition 4.20 (Normalized Discrete Surface Ricci Flow).

$$\frac{du_i}{dt} = \bar{K} - K_i, \tag{4.10}$$

where $\bar{K}$ is the average vertex curvature, $\bar{K} = 2\pi\chi(\Sigma)/n$, and n is the number of vertices.

It is easy to see that the normalized Ricci flow (4.10) is the negative gradient flow of the energy,

$$E(\mathbf{u}) = \int \sum_{i=1}^{n} (K_i - \bar{K}) du_i = \int \sum_{i=1}^{n} K_i du_i - \bar{K} \sum_{i=1}^{n} u_i.$$

If we ignore the linear term, then the energy is convex. It has a unique global minimum, where the gradient is zero, namely, $\nabla E(\mathbf{u}) = (K_1 - \bar{K}, K_2 - \bar{K}, \ldots, K_n - \bar{K})$ is a zero vector. If the average curvature is in the admissible curvature space, then the negative gradient flow leads to the global minimum. Therefore, we show the convergence theorem. In order to check whether the average curvature is admissible, we can verify the inequality (4.18) for all proper vertex subsets.

Theorem 4.6 (Convergence of Discrete Ricci Flow). *Suppose Σ is a triangular mesh with nonobtuse intersection angles $\phi_{ij} \in [0, \frac{\pi}{2}]$. If for any proper subset of the vertices $I \subset V$,*

$$\frac{2\pi|I|\chi(\Sigma)}{n} > - \sum_{(e,v)\in Lk(I)} (\pi - \phi(e)) + 2\pi\chi(F_I),$$

then the discrete surface Ricci flow converges to the metric of constant curvature $2\pi(\Sigma)/n$.

Furthermore, we can estimate the convergence rate of the discrete Ricci flow.

Theorem 4.7 (Exponential Convergence Rate). *The discrete surface Ricci flow converges exponentially fast to the constant curvature metric, for every vertex v_i,*

$$(K_i(t) - \bar{K})^2 \le c_2 e^{-c_1 t},$$

where c_1, c_2 are positive constants.

In order to prove the convergence and convergence rate of the discrete Ricci flow, we need the following lemma.

Lemma 4.11. *Suppose Σ is a connected triangular mesh. Then there exists a constant $c_3 > 0$ depending only on Σ, so that*

$$\sum_{i=1}^{n}(K_i - \bar{K})^2 \le c_3 \sum_{[v_i,v_j]\in\Sigma} (K_i - K_j)^2, \tag{4.11}$$

for all time $t > 0$.

Proof. According to the Cauchy's inequality $\langle \mathbf{x}, \mathbf{y} \rangle^2 \le \langle \mathbf{x}, \mathbf{x} \rangle \langle \mathbf{y}, \mathbf{y} \rangle$, let

$$\mathbf{x} = (a_i - a_1, a_i - a_2, \ldots, a_i - a_n)^T, \mathbf{y} = (1, 1, \ldots, 1),$$

then

$$(a_i - \bar{a})^2 \le \frac{1}{n}\left(\sum_{j=1}^{n}(a_i - a_j)^2\right),$$

where $\bar{a} = \frac{\sum_{i=1}^{n} a_i}{n}$. We obtain

$$\sum_{i=1}^{n}(K_i - \bar{K})^2 \le \frac{1}{n}\sum_{i,j=1}^{n}(K_i - K_j)^2. \tag{4.12}$$

On the other hand, since the surface Σ is connected, for any two vertices $v_i, v_j \in \Sigma$, there exists a shortest path between them. The path is represented as a sequence of vertices $v_{m_1} = v_i, \ldots, v_{m_l} = v_j$, so that v_{m_k} and $v_{m_{k+1}}$ are adjacent. The value

$$(K_i - K_j)^2 = \left(\sum_{k=1}^{l-1}(K_{m_k} - K_{m_{k+1}})\right)^2 \le \sum_{k=1}^{l-1}(K_{m_k} - K_{m_{k+1}})^2.$$

Therefore

$$\sum_{i,j=1}^{n}(K_i - K_j)^2 \le c_3 n \sum_{[v_i,v_j]\in\Sigma}(K_i - K_j)^2. \tag{4.13}$$

By combining (4.12) and (4.13), we obtain

$$\sum_{i=1}^{n}(K_i - \bar{K})^2 \le c_3 \sum_{[v_i,v_j]\in\Sigma} (K_i - K_j)^2.$$

This completes the proof for the lemma.   □

Now we prove the convergence theorems for the discrete Ricci flow.

Proof. During the flow, the energy $E(\mathbf{u})$ decreases monotonously. The path $\mathbf{u}(t)$ in the admissible metric space is contained in a compact subset, $\{\mathbf{u}|E(\mathbf{u}) \le E_0\}$, where E_0 is the initial energy value. So all $u_i(t)$'s are bounded. The edge weights w_{ij}'s are smooth functions of $\mathbf{u}$, therefore there exists a constant $c_4 > 0$, such that for all $[v_i, v_j] \in \Sigma$,

$$w_{ij}(t) > c_4, t \in [0, \infty). \tag{4.14}$$

Consider the energy

$$g(t) := \sum_{i=1}^{n}(K_i(t) - \bar{K})^2.$$

From $n\bar{K} = \sum_{i=1}^{n} K_i(t)$,

$$g(t) = \sum_{i=1}^{n}(K_i^2+\bar{K}^2-2K_i\bar{K})=\sum_{i=1}^{n} K_i^2+n\bar{K}^2-\left(2\sum_{i=1}^{n} K_i\right)\bar{K} = \sum_{i=1}^{n} K_i^2(t)-n\bar{K}^2.$$

From $K_i' = \sum_j w_{ij}(u_i' - u_j') = \sum_j w_{ij}(K_j - K_i)$,

$$g'(t) := 2\sum_{i=1}^{n}\sum_{[v_i,v_j]\in\Sigma} w_{ij} K_i(K_j - K_i).$$

By switching the order of i, j, we also have

$$g'(t) = 2\sum_{j=1}^{n}\sum_{[v_i,v_j]\in\Sigma} w_{ij} K_j(K_i - K_j).$$

Thus from (4.14) and (4.11),

$$g'(t) = -\sum_{[v_i,v_j]\in\Sigma} w_{ij}(K_i - K_j)^2 \le -c_4 \sum_{[v_i,v_j]\in\Sigma} (K_i - K_j)^2 \le -\frac{c_4}{c_3}\sum_{i=1}^{n}(K_i - \bar{K})^2$$

$$= -\frac{c_4}{c_3}g(t).$$

This implies that $g(t) \leq c_2 e^{-\frac{c_4}{c_3}t}$ for all time $t \geq 0$ and for some constant c_2. Then we obtain

$$(K_i(t) - \bar{K})^2 \leq \sum_{i=1}^{n} (K_i - \bar{K})^2 = g(t) \leq c_2 e^{-\frac{c_4}{c_3}t}.$$

Let c_1 be $\frac{c_4}{c_3}$. This completes the proof. $\square$

4.2.7   *Unified Euclidean Discrete Surface Ricci Flow*

Thurston's circle packing has been generalized to different schemes. This greatly increases the flexibility for practical purposes. All the schemes share the same geometric principles. Therefore, a unified theory can be established to cover all the schemes.

Generalized Schemes

Figure 4.15 illustrates the general schemes, where the power circles are in *red* color. Frame (**a**) shows the classical Thurston's circle packing, where the edge length and the circle radii satisfy the relation

$$l_k^2 = \gamma_i^2 + \gamma_j^2 + 2\gamma_i \gamma_j \cos \phi_k.$$

where ϕ_k is the intersection angle between circles at v_i and v_j. Algebraically, this formula is generalized to the following form

$$l_k^2 = \alpha_i \gamma_i^2 + \alpha_j \gamma_j^2 + 2\gamma_i \gamma_j \eta_k, \tag{4.15}$$

where

$$\alpha_i, \alpha_j \in \{+1, 0, -1\}, \eta_k \geq 0.$$

Frame (**b**) shows the tangential circle packing, where the circle intersection angles become zeros, namely, all circles are tangential to each other, $\phi_k = 0$. Frame (**c**) shows the inversive distance circle packing, where all the circles are disjoint, the cosine of ϕ_k is replaced by the η_k, the so-called *inversive distance*, which is greater than 1. Frame (**d**) shows the discrete Yamabe flow scheme, where all α_i's are 0's. Frame (**e**) is the *imaginary radius circle packing*, namely, all radii are imaginary numbers, and α_i's equal -1.

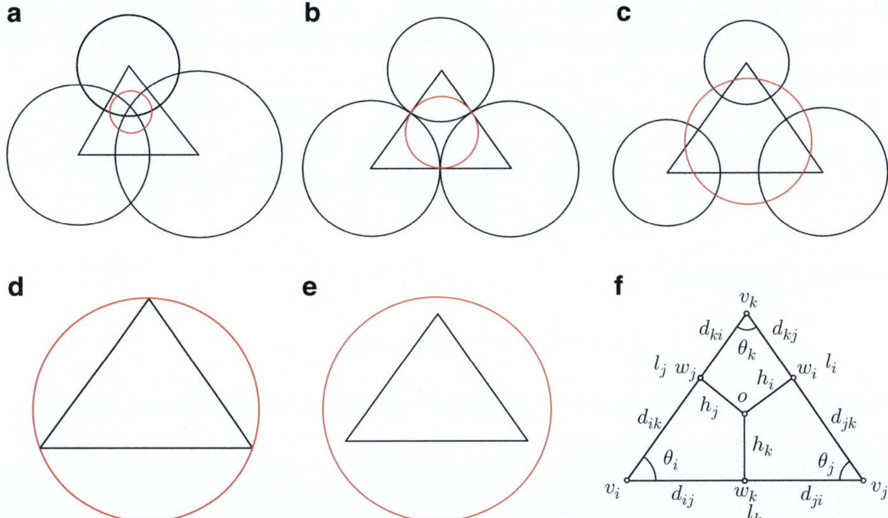

Fig. 4.15 Generalized schemes (**a**) Thurston's circle packing: $l_k^2 = \gamma_i^2 + \gamma_j^2 + 2\gamma_i\gamma_j \cos\phi_k$, (**b**) Tangential circle packing: $l_k = \gamma_i + \gamma_j$, (**c**) Inversive distance circle packing: $l_k^2 = \gamma_i^2 + \gamma_j^2 + 2\gamma_i\gamma_j\eta_k$, (**d**) Discrete Yamabe flow: $l_k^2 = 2\gamma_i\gamma_j\eta_k$, (**e**) Imaginary radius circle packing: $l_k^2 = -\gamma_i^2 - \gamma_j^2 + 2\gamma_i\gamma_j\eta_k$, (**f**) Common configuration

We define the *generalized power circle* and the *power center*. For the schemes on the top row, **a–c** the power circle is a circle orthogonal to three circles centered at the vertices, and its center is the power center. For the discrete Yamabe flow scheme **d**, the power circle is the circumcircle and the power center is the circumcenter. The power circle for the imaginary radius circle packing **e** is more complicated. Suppose the circle radii at v_i, v_j, v_k are $\gamma_i, \gamma_j, \gamma_k$, respectively. We translate v_i, v_j, v_k vertically (perpendicularly to the plane containing the triangle) to p_i, p_j, p_k, respectively, such that the vertical distances between the corresponding pairs are γ_i^2, γ_j^2 and γ_k^2, respectively. We draw a hemisphere S through $p_i, p_j,$ and p_k, and its equator is on the same plane containing the triangle. The equator of S is the power circle, and the spherical center is the power center. Algebraically, all the power centers satisfies the following *equidistance* condition,

$$|v_i - o|^2 - \alpha_i\gamma_i^2 = |v_j - o|^2 - \alpha_j\gamma_j^2 = |v_k - o|^2 - \alpha_k\gamma_k^2,$$

where o is the power center.

All the schemes share the same configuration as illustrated in frame **f**. We draw perpendicular lines from the power center to each edge. The distances from the power center to each edge are $h_i, h_j,$ and h_k, respectively. The pedal points divide each edge into two segments. Suppose the pedal point on $[v_i, v_j]$ is w_k, the length of $[v_i, w_k]$ is d_{ij}, and the length of $[v_j, w_k]$ is d_{ji}. We can show all the key formulae for Thurston's circle packing hold for the generalized schemes. From the following equations,

$$d_{ij}^2 + h_k^2 = d_{ik}^2 + h_j^2 = |v_i - o|^2,$$
$$d_{jk}^2 + h_i^2 = d_{ji}^2 + h_k^2 = |v_j - o|^2,$$
$$d_{ki}^2 + h_j^2 = d_{kj}^2 + h_i^2 = |v_k - o|^2,$$

we obtain the relation

$$d_{ij}^2 + d_{jk}^2 + d_{ki}^2 = d_{ji}^2 + d_{ik}^2 + d_{kj}^2.$$

The differential relation between edge lengths and the conformal factors also holds,

$$\frac{\partial l_k}{\partial u_i} = d_{ij}, \quad \frac{\partial l_k}{\partial u_j} = d_{ji},$$
$$\frac{\partial l_i}{\partial u_j} = d_{jk}, \quad \frac{\partial l_i}{\partial u_k} = d_{kj},$$
$$\frac{\partial l_j}{\partial u_i} = d_{ik}, \quad \frac{\partial l_j}{\partial u_k} = d_{ki}.$$

Unified Theorems

All the schemes can be unified as the discrete conformal structure [9].

Definition 4.21 (Discrete Conformal Structure). A triangular mesh Σ admits a discrete conformal structure if it admits a length function $l : E \to \mathbb{R}$, such that

1. $l_{ij}(u_i, u_j)$ is symmetric in u_i and u_j, where $u : V \to \mathbb{R}$ is an arbitrary function defined on V.
2. $l_{ij} = d_{ij} + d_{ji}$, where

$$\frac{\partial l_{ij}}{\partial u_i} = d_{ij}.$$

3. For any triangle $[v_i, v_j, v_k]$,

$$d_{ij}^2 + d_{jk}^2 + d_{ki}^2 = d_{ji}^2 + d_{kj}^2 + d_{ik}^2.$$

The concept of discrete conformal structure is more general, different schemes can be combined together, as shown in Fig. 4.16, where the inversive distance circle packing (at v_i), discrete Yamabe flow (at v_k), and imaginary radius circle packing (v_j) schemes are mixed together. The edge lengths are given by

$$l_k^2 = \gamma_i^2 - \gamma_j^2 + 2\gamma_i \gamma_j \eta_k, \ l_j^2 = \gamma_i^2 + 2\gamma_i \gamma_k \eta_j, \ l_i^2 = -\gamma_j^2 + 2\gamma_j \gamma_k \eta_i.$$

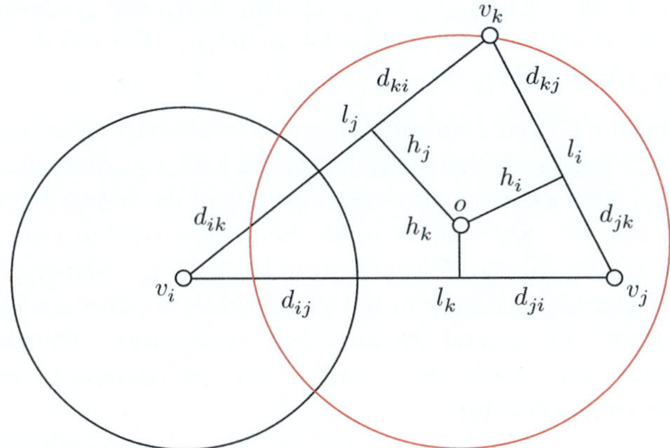

Fig. 4.16 Mixed scheme

With a general discrete conformal structure, the following theoretical results hold. The differential relations between the inner angles and the conformal factors are given by

$$\frac{\partial \theta_i}{\partial u_j} = \frac{\partial \theta_j}{\partial u_i} = \frac{h_k}{l_k}, \quad \frac{\partial \theta_j}{\partial u_k} = \frac{\partial \theta_k}{\partial u_j} = \frac{h_i}{l_i}, \quad \frac{\partial \theta_k}{\partial u_i} = \frac{\partial \theta_i}{\partial u_k} = \frac{h_j}{l_j}.$$

Also, the following formulae hold for all the schemes,

$$\frac{\partial \theta_i}{\partial u_i} = -\frac{\partial \theta_i}{\partial u_j} - \frac{\partial \theta_i}{\partial u_k}, \quad \frac{\partial \theta_j}{\partial u_j} = -\frac{\partial \theta_j}{\partial u_k} - \frac{\partial \theta_j}{\partial u_i}, \quad \frac{\partial \theta_k}{\partial u_k} = -\frac{\partial \theta_k}{\partial u_i} - \frac{\partial \theta_k}{\partial u_j}.$$

These equations imply the following lemma.

Lemma 4.12. *Given a triangle $[v_i, v_j, v_k]$ with a discrete conformal structure, assume three discrete conformal factors are u_i, u_j, u_k, and the three inner angles are $\theta_k, \theta_i, \theta_j$. The 1-form*

$$\omega = \theta_i d u_i + \theta_j d u_j + \theta_k d u_k$$

is a closed 1-form.

Then we can define discrete Ricci energy on one triangle.

Definition 4.22 (Discrete Ricci Energy). Given a triangle $[v_i, v_j, v_k]$ with a discrete conformal structure, assume three discrete conformal factors are u_i, u_j, u_k, and the three inner angles are $\theta_k, \theta_i, \theta_j$. The discrete Ricci energy is defined as

$$E(u_i, u_j, u_k) = \int_{(0,0,0)}^{(u_i, u_j, u_k)} \omega.$$

The following theorem holds for a triangle with a discrete conformal structure.

Theorem 4.8. *Given a triangle $[v_i, v_j, v_k]$ with a discrete conformal structure, assume three discrete conformal factors are u_i, u_j, u_k. The discrete Ricci energy is convex on the plane $u_i + u_j + u_k = 0$.*

We can define the discrete conformal structure for a triangular mesh Σ and define the discrete Ricci energy on the whole mesh. Then we can show the convexity of the Ricci energy, but we cannot show the convexity of the admissible metric space. Therefore, we can get local rigidity result, not global rigidity (see Theorem 4.5 for the global rigidity of the Thurston's circle packing). Namely, the mapping from a discrete conformal metric to the discrete Gauss curvature is locally one-to-one. Furthermore, for general schemes, there is no explicit method to describe the admissible curvature space (see Lemma 4.10 for the admissible curvature space of the Thurton's circle packing).

Geometric Interpretation of Ricci Energy

The discrete Ricci energy on a triangle with a conformal structure has geometric interpretation, which is the volume of a hyperbolic tetrahedron. The upper half space model of three-dimensional hyperbolic space $\mathbb{H}^3$ is the space

$$\mathbb{H}^3 = \{(x, y, z) \in \mathbb{R}^3 | z > 0\}$$

with the Riemannian metric

$$\frac{dx^2 + dy^2 + dz^2}{z^2}.$$

The xy-plane is the plane at infinity. Geodesics are circular arcs perpendicular to the xy-plane. Hyperbolic planes are hemispheres, whose equators are on the xy-plane. An ideal tetrahedron has four vertices at infinity, namely, on xy-plane. If some vertices exceed the xy-plane, then the tetrahedron is called a truncated ideal tetrahedron. Basically, a triangle with a conformal structure corresponds to a truncated ideal hyperbolic tetrahedron in $\mathbb{H}^3$, and the discrete Ricci energy equals the volume of the hyperbolic tetrahedron up to a constant. In the following, we visualize the Ricci energy for different schemes.

The geometric meanings of discrete Ricci energy for Thurston's circle packing, tangential circle packing, and inversive distance circle packing are illustrated in Figs. 4.17, 4.18, and 4.19, respectively. The truncated hyperbolic ideal tetrahedra are constructed as follows. First, we construct a prism in $\mathbb{H}^3$, which is the direct product of the triangle $[v_i, v_j, v_k]$ with the vertical line segment $[0, c]$, where $c > 0$ is sufficiently large. Through each vertex circle and the power circle, we construct a hemisphere. Each hemisphere is a hyperbolic plane. We cut off the prism by these four hyperbolic planes to obtain the truncated idea tetrahedron.

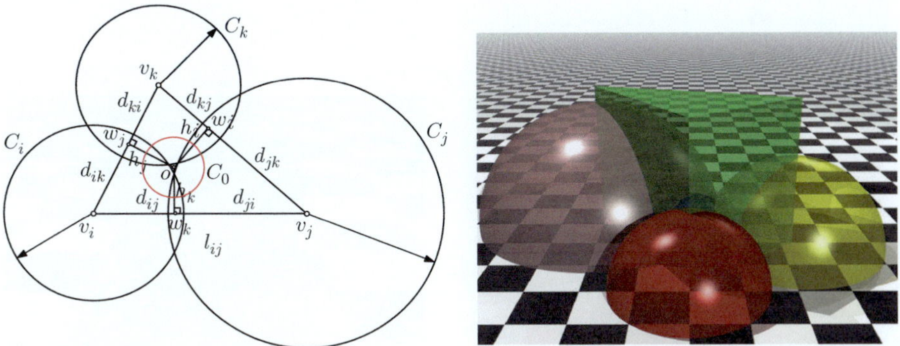

Fig. 4.17 Thurston's circle packing scheme

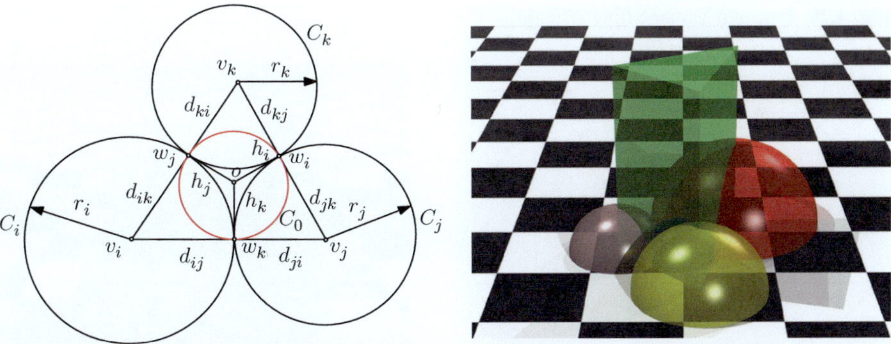

Fig. 4.18 Tangential circle packing scheme

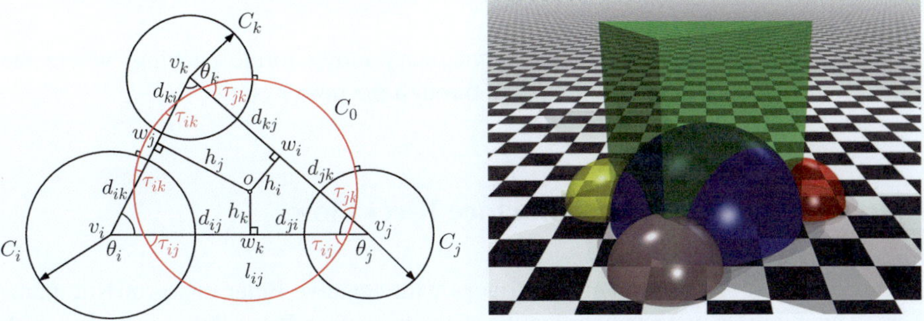

Fig. 4.19 Inversive distance circle packing scheme

Figure 4.20 visualizes the Ricci energy for discrete Yamabe flow scheme. The prism is cut off by the hyperbolic plane through the power circle. Then for each vertex v_i, we construct a sphere tangent to v_i with (Euclidean) radius r_i. Then the tetrahedron is cut off by the spheres. The volume of the resulting tetrahedron is the discrete Ricci energy up to a constant.

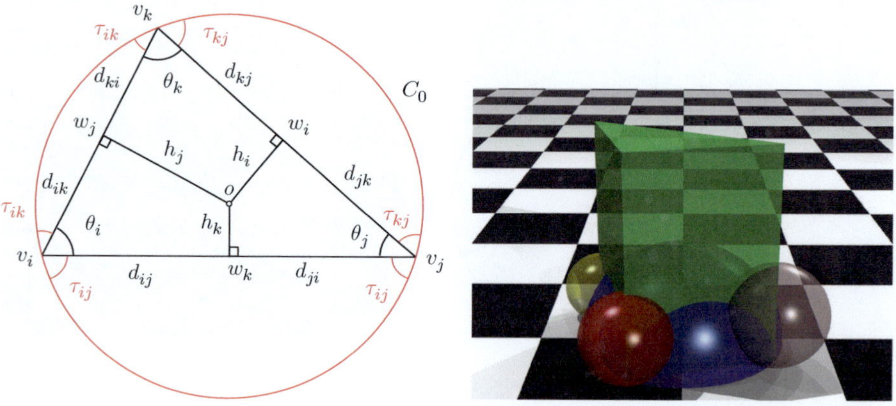

Fig. 4.20 Discrete Yamabe flow scheme

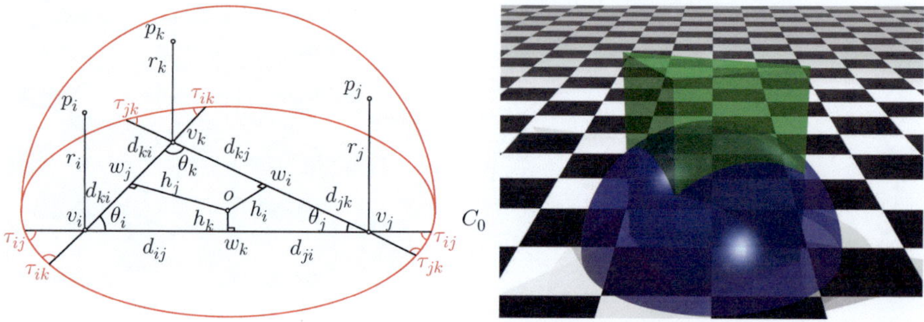

Fig. 4.21 Imaginary radius circle packing scheme

Figure 4.21 explains the case for imaginary radius circle packing, where the prism is cut off by the hyperbolic plane through the power circle.

4.3   Hyperbolic Discrete Surface Ricci Flow

In practice, it is crucial to handle surfaces with negative Euler characteristic numbers. This requires us to generalize the discrete surface Ricci flow to meshes with hyperbolic background geometry. In the following, we generalize the theoretical frameworks to hyperbolic discrete surfaces. Figure 4.22 shows the uniformization of a genus two surface, which is computed using the discrete hyperbolic Ricci flow method.

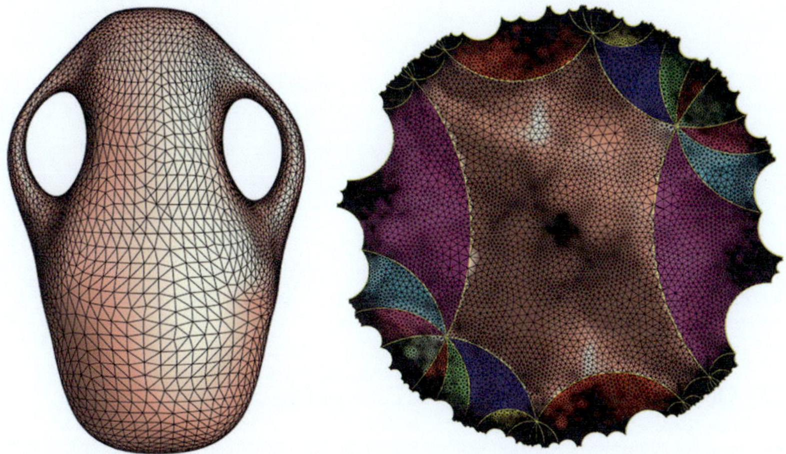

Fig. 4.22 Uniformization of a high genus surface using hyperbolic Ricci flow

4.3.1   Hyperbolic Derivative Cosine Law

The hyperbolic cosine law and sine law are given by

$$\cos \theta_i = \frac{\cosh l_j \cosh l_k - \cosh l_i}{\sinh l_j \sinh l_k},$$

$$\frac{\sin \theta_i}{\sinh l_i} = \frac{\sin \theta_j}{\sinh l_j} = \frac{\sin \theta_k}{\sinh l_k}.$$

The double area of the hyperbolic triangle is

$$A = \sinh l_j \sinh l_k \sin \theta_i.$$

Figure 4.23 shows a hyperbolic triangle and the hyperbolic circle packing.

Lemma 4.13 (Hyperbolic Derivative Cosine Law).

$$\frac{\partial \theta_i}{\partial l_i} = \frac{\sinh l_i}{A}, \ \frac{\partial \theta_i}{\partial l_j} = -\frac{\sinh l_i}{A} \cos \theta_k.$$

Proof. From the hyperbolic cosine law,

$$\cos \theta_i = \frac{\cosh l_j \cosh l_k - \cosh l_i}{\sinh l_j \sinh l_k}.$$

Taking derivative on both sides with respect to l_i, we get

$$-\sin \theta_i \frac{\partial \theta_i}{\partial l_i} = \frac{-\sinh l_i}{\sinh l_j \sinh l_k},$$

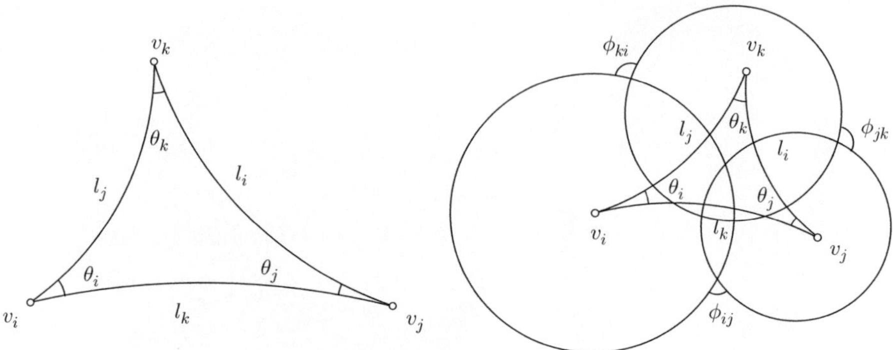

Fig. 4.23 A hyperbolic triangle and hyperbolic circle packing

therefore

$$\frac{\partial \theta_i}{\partial l_i} = \frac{\sinh l_i}{A}.$$

Similarly,

$$\cos \theta_i = \frac{\cosh l_j \cosh l_k - \cosh l_i}{\sinh l_j \sinh l_k}$$

$$\frac{\partial \theta_i}{\partial l_j} = \frac{\sinh l_j \cosh l_k (\sinh l_j \sinh l_k) - (-\cosh l_i + \cosh l_j \cosh l_k) \sinh l_k \cosh l_j}{-\sin \theta_i (\sinh l_j \sinh l_k)^2}$$

$$= \frac{(\sinh l_j^2 - \cosh l_j^2) \cosh l_k \sinh l_k + \cosh l_i \cosh l_j \sinh l_k}{-\sin \theta_i \sinh l_j^2 \sinh l_k^2}$$

$$= \frac{-\cosh l_k + \cosh l_i \cosh l_j}{-A \sinh l_j}$$

$$= \frac{\sinh l_i \sinh l_j \cos \theta_k}{-A \sinh l_j}$$

$$= -\frac{\sinh l_i}{A} \cos \theta_k.$$

$\square$

We write the result in the matrix format,

$$\begin{pmatrix} d\theta_i \\ d\theta_j \\ d\theta_k \end{pmatrix} = \frac{-1}{A} \begin{pmatrix} \sinh_i & 0 & 0 \\ 0 & \sinh_j & 0 \\ 0 & 0 & \sinh_k \end{pmatrix} \begin{pmatrix} -1 & \cos \theta_k & \cos \theta_j \\ \cos \theta_k & -1 & \cos \theta_i \\ \cos \theta_j & \cos \theta_i & -1 \end{pmatrix} \begin{pmatrix} dl_i \\ dl_j \\ dl_k \end{pmatrix}. \quad (4.16)$$

4.3.2 Thurston's Circle Packing

Similarly to the Euclidean case, we can use Thurston's circle packing for discrete conformal metric deformation. Then the relation between the edge length and the circle radii is governed by the hyperbolic cosine law,

$$\cosh l_i = \cosh r_j \cosh r_k + \sinh r_j \sinh r_k I_{jk},$$

where l_i is the length of the edge $[v_i, v_j]$, and I_{jk} is the cosine of the intersection angle between two circles centered at v_j and v_k.

Lemma 4.14. *The differential relation between the edge lengths and the circle radii is*

$$dl_i = \frac{-\cosh r_k + \cosh l_i \cosh r_j}{\sinh l_i \sinh r_j} dr_j + \frac{-\cosh r_j + \cosh l_i \cosh r_k}{\sinh l_i \sinh r_k} dr_k.$$

Proof.

$$\cosh l_i = \cosh r_j \cosh r_k + \sinh r_j \sinh r_k I_{jk},$$

$$\sinh l_i \frac{d l_i}{d r_j} = \sinh r_j \cosh r_k + \cosh r_j \sinh r_k I_{jk},$$

$$\frac{d l_i}{d r_j} = \frac{\sinh r_j \cosh r_k + \cosh r_j \sinh r_k I_{jk}}{\sinh l_i}.$$

Because

$$I_{jk} = \frac{\cosh l_i - \cosh r_j \cosh r_k}{\sinh r_j \sinh r_k}.$$

$$\frac{d l_i}{d r_j} = \frac{\sinh r_j \cosh r_k + \cosh r_j \sinh r_k \frac{\cosh l_i - \cosh r_j \cosh r_k}{\sinh r_j \sinh r_k}}{\sinh l_i}$$

$$= \frac{\sinh^2 r_j \cosh r_k + \cosh r_j \cosh l_i - \cosh^2 r_j \cosh r_k}{\sinh l_i \sinh r_j}$$

$$= \frac{(\sinh^2 r_j - \cosh^2 r_j) \cosh r_k + \cosh r_j \cosh l_i}{\sinh l_i \sinh r_j}$$

$$= \frac{\cosh r_j \cosh l_i - \cosh r_k}{\sinh l_i \sinh r_j}.$$

$\square$

We write the equation in matrix format,

$$
\begin{pmatrix} dl_i \\ dl_j \\ dl_k \end{pmatrix} = \begin{pmatrix} \frac{1}{\sinh l_i} & 0 & 0 \\ 0 & \frac{1}{\sinh l_j} & 0 \\ 0 & 0 & \frac{1}{\sinh l_k} \end{pmatrix} M \begin{pmatrix} \frac{1}{\sinh r_i} & 0 & 0 \\ 0 & \frac{1}{\sinh r_j} & 0 \\ 0 & 0 & \frac{1}{\sinh r_k} \end{pmatrix} \begin{pmatrix} dr_i \\ dr_j \\ dr_k \end{pmatrix}, \qquad (4.17)
$$

where

$$
M = \begin{pmatrix} 0 & -\cosh r_k + \cosh l_i \cosh r_j & -\cosh r_j + \cosh l_i \cosh r_k \\ -\cosh r_k + \cosh l_j \cosh r_i & 0 & -\cosh r_i + \cosh l_j \cosh r_k \\ -\cosh r_j + \cosh l_k \cosh r_i & -\cosh r_i + \cosh l_k \cosh r_j & 0 \end{pmatrix}.
$$

Let the discrete conformal factor $u_i = \log\tanh\frac{r_i}{2}$, $(a,b,c) = (\cosh l_i, \cosh l_j, \cosh l_k)$, and $(x,y,z) = (\cosh r_i, \cosh r_j, \cosh r_k)$. By the hyperbolic cosine law, we get the following lemma.

Lemma 4.15. *The differential relations between the inner angles and discrete conformal factors are*

$$
\begin{pmatrix} d\theta_i \\ d\theta_j \\ d\theta_k \end{pmatrix} = N \begin{pmatrix} du_i \\ du_j \\ du_k \end{pmatrix},
$$

where

$$
N = \begin{pmatrix} 1 - a^2 & ab - c & ca - b \\ ab - c & 1 - b^2 & bc - a \\ ca - b & bc - a & 1 - c^2 \end{pmatrix} \begin{pmatrix} \frac{1}{a^2 - 1} & 0 & 0 \\ 0 & \frac{1}{b^2 - 1} & 0 \\ 0 & 0 & \frac{1}{c^2 - 1} \end{pmatrix} \begin{pmatrix} 0 & ay - z & az - y \\ bx - z & 0 & bz - x \\ cx - y & cy - x & 0 \end{pmatrix}.
$$

Proof. From $u_i = \log\tanh\frac{r_i}{2}$, we obtain $dr_i = \sinh r_i\, du_i$. Combining it with (4.16) and (4.17), we get the formula. $\qquad\square$

Lemma 4.16. *The differential form $\omega = \theta_i\, du_i + \theta_j\, du_j + \theta_k\, du_k$ is closed.*

Proof. By direct computation, for example,

$$
\frac{\partial \theta_i}{\partial u_j} = \frac{\partial \theta_j}{\partial u_i} = z - \frac{ac - b}{c^2 - 1}x - \frac{bc - a}{c^2 - 1}y.
$$

$\qquad\square$

Lemma 4.17. *The admissible metric space is $\mathbb{R}^3$.*

Proof. Because $l_i \le r_j + r_k \le l_j + l_k$, the triangle inequality holds for all possible $\{r_i, r_j, r_k\}$. $\qquad\square$

Because the admissible metric space is simply connected, the closed 1-form is exact. We can define the discrete Ricci energy as $\int \omega$.

Lemma 4.18. *The discrete Ricci energy*

$$E(u_i, u_j, u_k) = \int_{(0,0,0)}^{(u_i, u_j, u_k)} \omega$$

is strictly concave.

Proof. It is sufficient to show the Hessian matrix $\left(\frac{\partial \theta_i}{\partial u_j}\right)$ is negative definite. Suppose we increase u_i. Then the hyperbolic triangle area increases as well. According to the Gauss–Bonnet theorem,

$$\pi - (\theta_i + \theta_j + \theta_k) = A,$$

therefore $\theta_i + \theta_j + \theta_k$ decreases,

$$\frac{\partial(\theta_i + \theta_j + \theta_k)}{\partial u_i} < 0.$$

This means

$$\frac{\partial \theta_i}{\partial u_i} < -\frac{\partial \theta_j}{\partial u_i} - \frac{\partial \theta_k}{\partial u_i} = -\frac{\partial \theta_i}{\partial u_j} - \frac{\partial \theta_i}{\partial u_k}.$$

On the other hand, $\frac{\partial \theta_i}{\partial u_j}$ is positive. So the negative of the Hessian matrix is diagonal dominant, and the Hessian matrix is negative definite. The energy is strictly concave. $\square$

4.3.3 Discrete Hyperbolic Ricci Energy

Let Σ be a triangular mesh with hyperbolic background geometry, and Φ be the angle intersection angles, such that for any edge $[v_i, v_j]$, $\phi_{ij} \in [0, \frac{\pi}{2}]$, and $u_i = \log \tanh \frac{r_i}{2}$. We can directly show that

$$\omega = \sum_{i=1}^{n} K_i d u_i$$

is a closed 1-form. The *admissible metric space* for all circle-packing metrics $\mathbf{u} = (u_1, u_2, \ldots, u_n)$ of (Σ, Φ) is $\mathbb{R}^n$. The discrete Ricci energy $\int \omega$ is strictly convex. The curvature map $\kappa : (u_1, u_2, \ldots, u_n) \to (K_1, K_2, \ldots, K_n)$ is a global diffeomorphism.

Let I be a proper subset of vertices V. F_I is the collection of faces, whose vertices are in I. The link of I, $Lk(I)$ is the set of pairs (e, v), where e is an edge whose vertices are not in I, and vertex $v \in I$, so that (e, v) forms a triangle. Given a curvature vector $(K_1, K_2, \ldots, K_n)$, if for any proper subset $I \subset V$, the following inequality (4.18) holds,

$$\sum_{v_i \in I} K_i > - \sum_{(e,v) \in Lk(I)} (\pi - \phi(e)) + 2\pi \chi(F_I), \tag{4.18}$$

then the curvature is admissible.

4.3.4 Generalized Schemes

Similarly to Euclidean case, Thurston's circle packing can be generalized to different schemes for discrete hyperbolic Ricci flow. In the following, we only discuss hyperbolic Yamabe flow; other schemes can be carried out similarly. Hyperbolic Yamabe flow is defined as

$$\sinh \frac{y_k}{2} = e^{u_i} \sinh \frac{l_k}{2} e^{u_j},$$

where $y_k(t)$ is the edge length during the flow, l_k is the initial edge length, and $y_k(0) = l_k$. Taking the derivative with respect to u_i on both sides,

$$\frac{1}{2} \cosh \frac{y_k}{2} \frac{\partial y_k}{\partial u_i} = e^{u_i} \sinh \frac{l_k}{2} e^{u_j} = \sinh \frac{y_k}{2},$$

we get

$$\frac{\partial y_k}{\partial u_i} = 2 \tanh \frac{y_k}{2} = 2 \frac{\sinh \frac{y_k}{2}}{\cosh \frac{y_k}{2}} = \frac{2 \sinh \frac{y_k}{2} \cosh \frac{y_k}{2}}{\cosh^2 \frac{y_k}{2}}.$$

Since $\sinh 2x = 2 \sinh x \cosh x$ and $\cosh 2x = 2 \cosh^2 x - 1$, we obtain

$$\frac{\partial y_k}{\partial u_i} = \frac{2 \sinh y_k}{\cosh y_k + 1}.$$

Therefore

$$\begin{pmatrix} dy_1 \\ dy_2 \\ dy_3 \end{pmatrix} = 2 \begin{pmatrix} \frac{\sinh y_1}{\cosh y_1 + 1} & 0 & 0 \\ 0 & \frac{\sinh y_2}{\cosh y_2 + 1} & 0 \\ 0 & 0 & \frac{\sinh y_3}{\cosh y_3 + 1} \end{pmatrix} \begin{pmatrix} 0 & 1 & 1 \\ 1 & 0 & 1 \\ 1 & 1 & 0 \end{pmatrix} \begin{pmatrix} du_1 \\ du_2 \\ du_3 \end{pmatrix}.$$

We use C_i, S_j to represent $\cosh y_i$ and $\sinh y_j$, respectively, and obtain

$$
\begin{pmatrix} d\theta_1 \\ d\theta_2 \\ d\theta_3 \end{pmatrix} = -\frac{2}{A}
\begin{pmatrix} S_1 & 0 & 0 \\ 0 & S_2 & 0 \\ 0 & 0 & S_3 \end{pmatrix}
\begin{pmatrix} -1 & \cos\theta_3 & \cos\theta_2 \\ \cos\theta_3 & -1 & \cos\theta_1 \\ \cos\theta_2 & \cos\theta_1 & -1 \end{pmatrix}
\begin{pmatrix} 0 & \frac{S_1}{C_1+1} & \frac{S_1}{C_1+1} \\ \frac{S_2}{C_2+1} & 0 & \frac{S_2}{C_2+1} \\ \frac{S_3}{C_3+1} & \frac{S_3}{C_3+1} & 0 \end{pmatrix}
\begin{pmatrix} du_1 \\ du_2 \\ du_3 \end{pmatrix}
$$

$$
= -\frac{2}{A}
\begin{pmatrix}
\frac{S_1 S_2 \cos\theta_3}{C_2+1} + \frac{S_1 S_3 \cos\theta_2}{C_3+1} & \frac{-S_1^2}{C_1+1} + \frac{S_1 S_3 \cos\theta_2}{C_3+1} & \frac{-S_1^2}{C_1+1} + \frac{S_1 S_2 \cos\theta_3}{C_2+1} \\
\frac{-S_2^2}{C_2+1} + \frac{S_2 S_3 \cos\theta_1}{C_3+1} & \frac{S_1 S_2 \cos\theta_3}{C_1+1} + \frac{S_2 S_3 \cos\theta_1}{C_3+1} & \frac{-S_2^2}{C_2+1} + \frac{S_1 S_2 \cos\theta_3}{C_1+1} \\
\frac{-S_3^2}{C_3+1} + \frac{S_2 S_3 \cos\theta_1}{C_2+1} & \frac{-S_3^2}{C_3+1} + \frac{S_1 S_3 \cos\theta_2}{C_1+1} & \frac{S_1 S_3 \cos\theta_2}{C_1+1} + \frac{S_2 S_3 \cos\theta_1}{C_2+1}
\end{pmatrix}
\begin{pmatrix} du_1 \\ du_2 \\ du_3 \end{pmatrix}.
$$

With the hyperbolic cosine law,

$$
\cosh y_i = \cosh y_j \cosh y_k - \sinh y_j \sinh y_k \cos\theta_i ,
$$

we get

$$
S_j S_k \cos\theta_i l = C_j C_k - C_i ,
$$

and $\cosh^2 x - \sinh^2 x = 1$. The right-hand side becomes

$$
-\frac{2}{A}
\begin{pmatrix}
\frac{C_1 C_2 - C_3}{C_2+1} + \frac{C_1 C_3 - C_2}{C_3+1} & \frac{1-C_1^2}{C_1+1} + \frac{C_1 C_3 - C_2}{C_3+1} & \frac{1-C_1^2}{C_1+1} + \frac{C_1 C_2 - C_3}{C_2+1} \\
\frac{1-C_2^2}{C_2+1} + \frac{C_2 C_3 - C_1}{C_3+1} & \frac{C_1 C_2 - C_3}{C_1+1} + \frac{C_2 C_3 - C_1}{C_3+1} & \frac{1-C_2^2}{C_2+1} + \frac{C_1 C_2 - C_3}{C_1+1} \\
\frac{1-C_3^2}{C_3+1} + \frac{C_2 C_3 - C_1}{C_2+1} & \frac{1-C_3^2}{C_3+1} + \frac{C_1 C_3 - C_2}{C_1+1} & \frac{C_1 C_3 - C_2}{C_1+1} + \frac{C_2 C_3 - C_1}{C_2+1}
\end{pmatrix}
\begin{pmatrix} du_1 \\ du_2 \\ du_3 \end{pmatrix}.
$$

We get

$$
\frac{\partial\theta_1}{\partial u_2} = -\frac{2}{A}\left(\frac{1-C_1^2}{C_1+1} + \frac{C_1 C_3 - C_2}{C_3+1} \right)
$$
$$
= -\frac{2}{A}\left(1 - C_1 + \frac{C_1 C_3 - C_2}{C_3+1} \right) = \frac{C_1 + C_2 - C_3 - 1}{A(1+C_3)/2}
$$

and

$$
\frac{\partial\theta_1}{\partial u_1} = -\frac{1}{A}\left(\frac{C_1 C_2 - C_3}{C_2+1} + \frac{C_1 C_3 - C_2}{C_3+1} \right)
$$
$$
= -\frac{2 C_1 C_2 C_3 - C_2^2 - C_3^2 + C_1 C_2 + C_1 C_3 - C_2 - C_3}{A(C_2+1)(C_3+1)/2}.
$$

By symmetry, we obtain

$$
\frac{\partial\theta_i}{\partial u_j} = \frac{\partial\theta_j}{\partial u_i} = \frac{C_i + C_j - C_k - 1}{A(C_k+1)/2}
$$

and

$$
\frac{\partial\theta_i}{\partial u_i} = -\frac{2 C_i C_j C_k - C_j^2 - C_k^2 + C_i C_j + C_i C_k - C_j - C_k}{A(C_j+1)(C_k+1)/2}.
$$

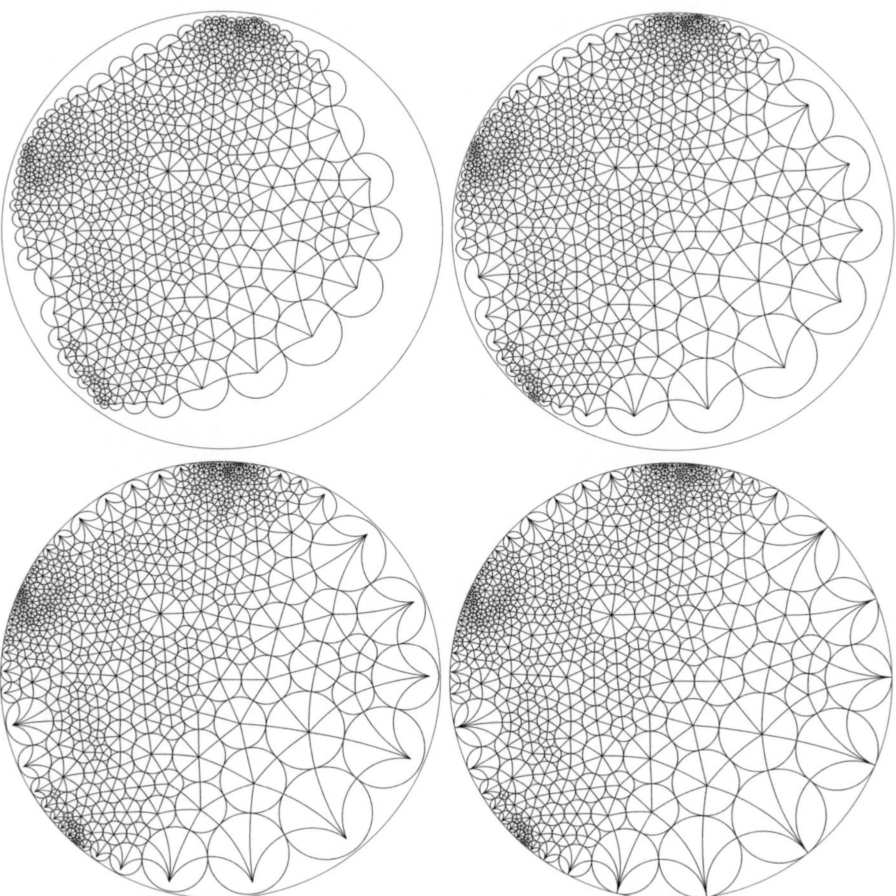

Fig. 4.24 Hyperbolic Ricci flow for Riemann mapping

Hyperbolic Ricci flow can also be applied for computing Riemann mapping. Set the target curvature to be zeros on the interior vertices, and the boundary circle radii go to ∞, then the Ricci flow leads to the Riemann mapping, as shown in Fig. 4.24.

Further Readings

The rigidity for classical circle packing was proved by Thurston [25], Marden–Rodin [19], Colin de Verdiére [26], Chow–Luo [6], Stephenson [24], and He [13]. Bowers–Stephenson [4] introduced inversive distance circle packing which generalizes Andreev–Thurston's intersection angle circle packing. See the work of Stephenson [24] for more information. Guo gave a proof for local rigidity [11]. Luo studied the combinatorial Yamabe problem for piecewise flat metrics on triangulated surfaces [16]. Springborn, Schröder, and Pinkall [23] considered this combinatorial conformal change of piecewise flat metrics and found an explicit formula of the energy function. Glickenstein [7,8] studied the combinatorial Yamabe flow on three-dimensional piecewise flat manifolds. Recently Glickenstein [9] set the theory of combinatorial Yamabe flow of piecewise flat metric in a broader context including the theory of circle packing on surfaces. Combinatorial Yamabe flow on hyperbolic surfaces with boundary has been studied by Guo in [10].

The variational approach to circle packing was first introduced by Colin de Verdiére [26]. Since then, many works on variational principles on circle packing or circle pattern have appeared. For example, see the works of Brägger [5], Rivin [20], Leibon [15], Chow–Luo [6], Bobenko–Springborn [3], Guo–Luo [12], and Springborn [22]. Variational principles for polyhedral surfaces including the topic of circle packing were studied systematically by Luo in [17]. Many energy functions are derived from the cosine law and its derivative. Tangential circle packing is generalized to tangential circle packing with a family of discrete curvatures. For exposition of this work, see also Luo–Gu–Dai [18].

References

1. Andreev, E.M.: Complex polyhedra in Lobachevsky spaces. (Russian) Mat. Sb. (N.S.) **81**(123), 445–478 (1970)
2. Andreev, E.M.: Convex polyhedra of finite volume in Lobachevsky space. (Russian) Mat. Sb. (N.S.) **83**(125), 256–260 (1970)
3. Bobenko, A.I., Springborn, B.A.: Variational principles for circle patterns and Koebe's theorem. Trans. Am. Math. Soc. **356**(2), 659–689 (2004)
4. Bowers, P.L., Stephenson, K.: Uniformizing dessins and Belyi maps via circle packing. Mem. Am. Math. Soc. **170**(805) (2004)
5. Brägger, W.: Kreispackungen und triangulierugen. Enseign. Math. **38**, 201–217 (1992)
6. Chow, B., Luo, F.: Combinatorial Ricci flows on surfaces. J. Different. Geomet. **63**(1), 97–129 (2003)
7. Glickenstein, D.: A combinatorial Yamabe flow in three dimensions. Topology **44**(4), 791–808 (2005)
8. Glickenstein, D.: A maximum principle for combinatorial Yamabe flow. Topology **44**(4), 809–825 (2005)
9. Glickenstein, D.: Discrete conformal variations and scalar curvature on piecewise flat two and three dimensional manifolds. J. Different. Geomet. **87**(2), 201–238 (2011)
10. Guo, R.: Combinatorial Yamabe flow on hyperbolic surfaces with boundary. Commun. Contemp. Mathe. **13**(5), 827–842 (2011)
11. Guo, R.: Local rigidity of inversive distance circle packing. Trans. Am. Math. Soc. **363**, 4757–4776 (2011)
12. Guo, R., Luo, F.: Rigidity of polyhedral surface ii. Geom. Topol. **13**, 1265–1312 (2009)
13. He, Z.X., Schramm, O.: On the convergence of circle packings to the Riemann map. Invent. Math. **125**, 285–305 (1996)
14. Koebe, P.: Kontaktprobleme der konformen abbildung. Ber. Sächs. Akad. Wiss. Leipzig, Math. Phys. Kl. **88**, 141–164 (1936)
15. Leibon, G.: Characterizing the Delaunay decompositions of compact hyperbolic surface. Geom. Topol. **6**, 361–391 (2002)

16. Luo, F.: Combinatorial Yamabe flow on surfaces. Contemp. Math. **6**(5), 765–780 (2004)
17. Luo, F.: Rigidity of polyhedral surfaces. arXiv:math.GT/0612714 (2006)
18. Luo, F., Gu, X., Dai, J.: Variational Principles for Discrete Surfaces. Advanced Lectures in Mathematics. High Education Press and International Press (2007)
19. Marden, A., Rodin, B.: Computational methods and function theory (Valpara'iso,1989). In: On Thurston's Formulation and Proof of Andreev's theorem. Lecture Notes in Mathematics, vol. 1435, pp. 103–116. Springer, Berlin (1990)
20. Rivin, I.: Euclidean structures of simplicial surfaces and hyperbolic volume. Ann. Math. **139**, 553–580 (1994)
21. Rodin, B., Sullivan, D.: The convergence of circle packings to the Riemann mapping. J. Differential Geomet. **26**, 349–360 (1987)
22. Springborn, B.: A variational principle for weighted Delaunay triangulation and hyperideal polyhedra. J. Different. Geomet. **78**(2), 333–367 (2008)
23. Springborn, B., Schröder, P., Pinkall, U.: Conformal equivalence of triangle meshes. ACM Trans. Graph. **27**(3), 1–11 (2008)
24. Stephenson, K.: Introduction to Circle Packing: The Theory of Discrete Analytic Functions. Cambridge University Press, Cambridge (2005)
25. Thurston, W.P.: The Geometry and Topology of 3-manifolds. Princeton University Press, Princeton (1981)
26. de Verdiére, Y.C.: Un principe variationnel pour les empilements de cercles. Invent. Math. **104**, 655–669 (1991)

Chapter 5
Algorithms and Applications

This chapter focuses on the computational algorithms and gives implementation details. In Chap. 4, we have discussed discrete Ricci flow based on different schemes. For simplicity, here we only give the algorithm based on the scheme of inversive distance circle packing. The algorithms for discrete surface Ricci flow based on all the other schemes are very similar. Also, we illustrate the details for the Euclidean discrete surface Ricci flow. The hyperbolic discrete surface Ricci flow algorithms are almost identical, except replacing the planar Euclidean geometry by hyperbolic geometry. Spherical surface mapping can be obtained through Euclidean Ricci flow, using the inverse stereographic projection of the planar map.

Furthermore, this chapter briefly introduces the direct applications of surface Ricci flow in computer vision and medical imaging. Two fundamental tasks are addressed in systematic ways:

1. *Surface registration and tracking*. This aims at finding a diffeomorphism with least distortions between surfaces with large deformations. We will cover different methods including harmonic mapping and quasi-conformal mapping.
2. *Shape analysis*. This aims at constructing a shape space, which is an infinite dimensional Riemannian manifold, to represent all the shapes, establish a Riemannian metric to measure the distance among shapes, and compute geodesics, which give the deformations among shapes.

All the algorithms are based on the theories discussed in previous chapters and generalized to discrete settings.

5.1 Discrete Surface Ricci Flow Algorithm

Suppose M is a triangular mesh embedded in $\mathbb{E}^3$ with the initial induced Euclidean metric and also with the prescribed target curvature $\bar{K} : V \to \mathbb{R}$. We are asked to compute a discrete Riemannian metric, which is conformal to the original metric

W. Zeng and X.D. Gu, *Ricci Flow for Shape Analysis and Surface Registration*, SpringerBriefs in Mathematics, DOI 10.1007/978-1-4614-8781-4__5, © Wei Zeng, Xianfeng David Gu 2013

Algorithm 5.1 Initial circle packing metric

Input: A triangular mesh M, embedded in $\mathbb{E}^3$.
Output: The initial circle packing metric.
1: **for all** face $[v_i, v_j, v_k] \in M$ **do**
2:     Compute $\gamma_i^{jk} = \frac{l_{ij} + l_{ki} - l_{jk}}{2}$.
3: **end for**
4: **for all** vertex $v_i \in M$ **do**
5:     Compute the radius $\gamma_i = \min_{jk} \gamma_i^{jk}$.
6: **end for**
7: **for all** edge $[v_i, v_j] \in M$ **do**
8:     Compute the inversive distance $\eta_{ij} = \frac{l_{ij}^2 - \gamma_i^2 - \gamma_j^2}{2\gamma_i \gamma_j}$.
9: **end for**

---

and realizes the target curvature. To address this problem, we will explain the computational algorithms for discrete surface Ricci flow [9,28]. In the following, we assume the meshes are with the Euclidean background geometry. The algorithms for meshes with the hyperbolic background geometry are almost identical, but require elementary knowledge about hyperbolic geometry.

In the first stage, the initial circle-packing metric is computed as illustrated in Algorithm 5.1. For simplicity, we explain the algorithm using inversive distance circle packing scheme. Other configurations are similar. For each vertex v_i, surrounded by faces $\{[v_i, v_j, v_k]\}$, we set the radius of the circle assigned to vertex v_i as

$$\gamma_i = \min_{jk} \frac{l_{ij} + l_{ki} - l_{jk}}{2}.$$

Since the mesh is with the Euclidean background geometry, for each edge $[v_i, v_j]$, we define the inversive distance between two circles (v_i, γ_i) and (v_j, γ_j) as

$$\eta_{ij} = \frac{l_{ij}^2 - \gamma_i^2 - \gamma_j^2}{2\gamma_i \gamma_j}. \tag{5.1}$$

In the second stage, we compute the desired metric for the prescribed target curvature by optimizing the Ricci energy $E(\mathbf{u}) = \int \sum_i (\bar{K}_i - K_i) du_i$, where $u_i = \log \gamma_i$ and $\mathbf{u} = (u_1, u_2, \ldots, u_n)^T$ (n is the number of vertices). Traverse all the faces, for each face $[v_i, v_j, v_k]$, compute the power circle that is orthogonal to three vertex circles and denote the power center as o_{ijk} and the distance from o_{ijk} to $[v_i, v_j]$ as h_{ij}^k. Go through all the edges, for each edge $[v_i, v_j]$, if it is adjacent to two faces $[v_i, v_j, v_k]$ and $[v_j, v_i, v_l]$, then its weight is given by

$$w_{ij} := \frac{h_{ij}^k + h_{ji}^l}{l_{ij}}. \tag{5.2}$$

Algorithm 5.2 Euclidean surface Ricci flow

Input: A triangular mesh M, the target curvature $\bar{K}$.
Output: A conformal Riemannian metric realizing $\bar{K}$.
  1: Compute the initial circle packing metric using Algorithm 5.1.
  2: **repeat**
  3:     For each face $[v_i, v_j, v_k]$, compute the power circle center o_{ijk}, compute the distances to three edges h_{ij}^k, h_{jk}^i and h_{ki}^j.
  4:     For each edge $[v_i, v_j]$, compute the edge weight w_{ij} using (5.2) and (5.3).
  5:     Construct the Hessian matrix, using (5.4).
  6:     Solve linear system $H\delta\mathbf{u} = \bar{K} - K$ restricted on $\sum_i u_i = 0$.
  7:     Update discrete conformal factor $\mathbf{u} \leftarrow \mathbf{u} + \delta\mathbf{u}$.
  8:     For each edge, compute the edge length $l_{ij}^2 = \gamma_i^2 + \gamma_j^2 + 2\gamma_i\gamma_i\eta_{ij}$.
  9:     For each face $[v_i, v_j, v_k]$, compute the corner angles $\theta_i^{jk}, \theta_j^{ki}$ and θ_k^{ij} using cosine law.
 10:     For each vertex v_i, compute the Gauss curvature $K_i = 2\pi - \sum_{jk} \theta_i^{jk}$.
 11: **until** $\max_{v_i \in M} |\bar{K}_i - K_i| < \epsilon$

If the edge only attaches to one face $[v_i, v_j, v_k]$, then

$$w_{ij} := \frac{h_{ij}^k}{l_{ij}}. \tag{5.3}$$

Then construct the Hessian matrix of the Ricci energy $H = (\frac{\partial^2 E}{\partial u_i \partial u_j})$, where

$$\frac{\partial^2 E}{\partial u_i \partial u_j} = \begin{cases} -w_{ij} & [v_i, v_j] \in M \\ \sum_k w_{ik} & i = j \\ 0 & \text{otherwise} \end{cases}. \tag{5.4}$$

Use Newton's method to optimize the energy $E(\mathbf{u})$ restricted on the hyper plane $\sum_i u_i = 0$,

$$\delta\mathbf{u} = H^{-1}\nabla E(\mathbf{u}) = H^{-1}(\bar{\mathbf{K}} - \mathbf{K}).$$

Algorithm 5.2 illustrates the pipeline of this stage.

In the final stage, we embed the mesh with the target metric onto the plane. We first compute the *cut graph* of the mesh as illustrated in Algorithm 5.3. Given a mesh M, construct its Poincaré dual mesh $\bar{M}$. Compute a spanning tree $\bar{T}$ of $\bar{M}$, which connects all the vertices in $\bar{M}$. Then we get a graph on the original mesh M,

$$G = \{e \in M | \bar{e} \notin \bar{T}\},$$

where $\bar{e} \in \bar{M}$ represents the dual edge of $e \in M$. Compute a spanning tree T of G, then

$$G - T = \{e_1, e_2, \ldots, e_k\},$$

Algorithm 5.3 Cut graph

Input: A triangular mesh M.
Output: A cut graph C of the mesh M.
  1: Compute the dual mesh $\bar{M}$ of M.
  2: Compute a spanning tree $\bar{T}$ of $\bar{M}$.
  3: Construct the graph $G = \{e \in M | \bar{e} \notin \bar{T}\}$.
  4: Compute a spanning tree T of G, $G - T = \{e_1, e_2, \ldots, e_k\}$.
  5: Each end vertex of v_i has a unique path to the root in the tree T, therefore $T \cup e_i$ has a unique loop γ_i.
  6: Return the cut graph C, which is the union of $\{\gamma_1, \gamma_2, \ldots, \gamma_k\}$.

Algorithm 5.4 Embedding

Input: A triangular mesh M.
Output: An isometric embedding ϕ.
  1: Compute a cut graph C of M using Algorithm 5.3.
  2: Slice M along the cut graph C to form a fundamental domain $\tilde{M}$.
  3: Embed a randomly chosen initial triangle $[v_0, v_1, v_2] \in \tilde{M}$ using (5.5).
  4: Put all the neighboring faces of the initial face to a queue.
  5: **while** the queue is not empty **do**
  6:     Pop the head face $[v_i, v_j, v_k]$ from the queue.
  7:     Suppose v_i and v_j has been embedded. Compute the intersection of two circles $(\phi(v_i), l_{ik}) \cap (\phi(v_j), l_{jk})$.
  8:     $\phi(v_k)$ is selected to keep the normal of the flattened face upward.
  9:     Put the neighboring faces of $[v_i, v_j, v_k]$, which haven't been accessed yet, to the queue.
10: **end while**

---

and $T \cup e_i$ has a unique loop γ_i. The *cut graph* C of the mesh M is given by

$$C := \bigcup_{i=1}^{k} \gamma_i.$$

The mesh M is sliced along the cut graph C to form a fundamental domain $\tilde{M}$. Then we compute the isometric embedding of $\tilde{M}$ onto the complex plane, $\phi : \tilde{M} \to \mathbb{C}$, with the target metric. First, randomly choose an initial face $[v_0, v_1, v_2]$, and isometrically flatten it on the plane,

$$\phi(v_0) = 0, \quad \phi(v_1) = l_{01}, \quad \phi(v_2) = l_{20}e^{i\theta_0^{12}}. \tag{5.5}$$

Then put all the neighboring faces to a queue. When the queue is not empty, popup the head face $[v_i, v_j, v_k]$. Assume $\phi(v_i)$ and $\phi(v_j)$ have been computed. Then $\phi(v_k)$ is at the intersection of two circles $(\phi(v_i), l_{ik})$ and $(\phi(v_j), l_{jk})$, furthermore, $\phi(v_k)$ is selected to keep the normal of the flattened face upward. Then put all the neighboring faces of $[v_i, v_j, v_k]$ to the queue, which haven't been flattened yet. Repeat flattening face by face until the queue is empty. Algorithm 5.4 illustrates the pipeline for the embedding process.

Hyperbolic discrete surface Ricci flow algorithm is very similar. Basically, all the computations are based on hyperbolic cosine law and carried out on the hyperbolic plane using Poincaré disk model. Many hyperbolic geometric operations can be converted to Euclidean operations. For example, because hyperbolic circles on the Poincaré disk coincide with Euclidean circles, finding the intersections between hyperbolic circles is equivalent to finding those between Euclidean circles.

Spherical surface mapping can be obtained by Euclidean Ricci flow. Given a genus zero closed mesh, we choose one triangle to be mapped to the north pole on the sphere and then remove that triangle to form a mesh with a boundary. Using the Euclidean Ricci flow to map the mesh onto the plane, such that the exterior angles at three boundary vertices are $\frac{2\pi}{3}$. Then we scale the planar image and use the inverse stereographic projection to map it onto the sphere. Finally, we fill the triangular hole at the north pole on the sphere.

5.2 Registration and Tracking

Shape registration and tracking are the fundamental tasks in many engineering fields, including computer vision and medical imaging. Surface registration aims at finding a diffeomorphic mapping between two given surfaces with Riemannian metrics $(S_k, \mathbf{g}_k), k = 1, 2, \phi : (S_1, \mathbf{g}_1) \to (S_2, \mathbf{g}_2)$, such that the mapping preserves some geometric properties or optimizes some special form of energy $E(\phi)$. Surface tracking is to register two successive surface frames in a sequence of surfaces with dynamic deformations.

Typical Types of Mappings

The followings are the types of mappings commonly pursued in practice.

Isometry. In medical imaging, human organs can be captured using computing tomography (CT) or magnetic resonance imaging (MRI) technologies. For example, the human brain cortical surfaces are often scanned and studied for diagnosis purposes. If the deformation between the two surfaces is small, such as the two scans of the same brain using different modalities, it is highly desirable to find an isometry between the two surfaces. Another example common in computer vision is clothes tracking. The deformations of clothes are almost isometric. The isometric mapping ϕ satisfies the following condition

$$\mathbf{g}_1 = \phi^* \mathbf{g}_2.$$

The detailed algorithm and application of isometry will be explained in Sect. 5.2.1.

Conformal Mapping. In virtual colonoscopy, human colon wall surfaces are scanned and reconstructed from CT images. The digitized surface is screened by human experts to locate the cancerous polyps. The colon wall surface has many folds, which can be unfolded onto the plane for better examination. This requires a mapping from the colon surface to the plane. It is mandatory for the mapping to preserve local shapes, such that the doctors can identify polyps on the planar image. Conformal mapping is a natural choice, which preserves the angles,

$$e^{2\lambda}\mathbf{g}_1 = \phi^*\mathbf{g}_2, \ \lambda : S_1 \to \mathbb{R},$$

where λ is the conformal factor. The algorithms and applications of conformal mapping will be explained in Sect. 5.2.1.

Optimal Mass Transport Map. Conformal mapping preserves angles but distorts area element. In many applications in medical imaging, it is desirable to preserve areas. For example, in brain morphology study, the cortical surfaces are mapped to the unit sphere. In order to measure the area of each functional region, it is required that the mapping preserves the area element. There are too many area preserving mappings, usually the one that optimizes the transport cost is applied. This can be formulated as follows. Suppose (S_1, μ_1) and (S_2, μ_2) are two planar domains with measures. The optimal mass transport map $\phi : S_1 \to S_2$ minimizes the following energy

$$E(\phi) := \int_{S_1} \|p - \phi(p)\|^2 d\mu_1(p),$$

with the area preserving constraint

$$\mu_1 = J(\phi)\mu_2,$$

where $J(\phi)$ is the Jacobian of ϕ. The optimal mass transport problem is equivalent to a Monge–Ampere equation, which can be solved using convex geometry [2].

Harmonic Map. In general, the deformation between two surfaces neither preserves angles nor areas, such as the deformation of human bladder, but the deformation is elastic, then the mapping will optimize the harmonic energy and can be formulated as harmonic maps. The harmonic energy solely depends on the conformal structure of the source and the Riemannian metric on the target. If the target surface has a Riemannian metric with negative curvature everywhere, then the degree one harmonic map is a diffeomorphism. Another advantage of harmonic map is that it can easily incorporate landmark constraints. The detailed algorithms and applications for harmonic maps will be explained in Sect. 5.2.2.

Quasi-Conformal Map. Suppose S_1 and S_2 are compact. Then general homeomorphisms between them can be modeled as quasi-conformal maps. Roughly speaking, each quasi-conformal map corresponds to its Beltrami differential. The mapping space of all the homeomorphism is converted to the space of Beltrami differentials. By manipulating the Beltrami differentials, we can optimize the mapping. For example, for human expression and large deformation of human organs, quasi-conformal mapping is a good modeling tool. The implementation details of the algorithms and application examples will be explained in Sect. 5.2.3.

Common Framework

All the methods for surface registration and tracking share the same framework. By mapping 3D surfaces to canonical planar domains using Ricci flow, 3D geometric processing tasks can be converted to image processing problems. This greatly improves the efficiency and efficacy of computational algorithms. The framework of surface matching and registration can be summarized as diagram (5.6).

$$
\begin{array}{ccc}
S_1 & \xrightarrow{\ f\ } & S_2 \\
f_1 \downarrow & & \downarrow f_2 \\
D_1 & \xrightarrow{\ \phi\ } & D_2
\end{array}
\tag{5.6}
$$

Suppose $S_k, k = 1, 2$ are the input *source* and *target* surfaces, respectively. In order to compute the optimal diffeomorphism $f : S_1 \to S_2$, we conformally map them onto the plane by $f_k : S_k \to D_k$, where D_k are called *conformal parameter domains*, then construct a planar diffeomorphism $\phi : D_1 \to D_2$. The registration between two surfaces is given by $f = f_2^{-1} \circ \phi \circ f_1$. Moreover, ϕ can be further optimized in the 2D mapping space. This provides a key to constructing the globally optimal diffeomorphism between surfaces.

5.2.1 Isometric and Conformal Mapping

Suppose two surfaces differ by a nearly isometric or conformal deformation, $f : S_1 \to S_2$, and are conformally mapped to the canonical domains, $f_k : S_k \to D_k$, $k = 1, 2$. Since f_1 and f_2 are conformal and the isometric mapping f is also conformal, $\phi = f_2 \circ f \circ f_1^{-1}$ is conformal. With appropriate normalization conditions, ϕ becomes the identity, namely, conformal images of isometric surfaces are identical on the 2D canonical domain.

Figure 5.1 demonstrates an example. Two surface scans S_1, S_2 come from the same plastic mask, but S_2 is with a bending deformation (no stretching). This deformation is very close to an isometry due to the plastic material property. Two surfaces are conformally mapped to circle domains D_1, D_2 by Ricci flow, where a Möbius normalization is applied to move the nose tip to the origin and rotate the middle point of the inner corners of the eyes to be on the imaginary axis. We can see that their images on D_1 and D_2 are almost identical. Therefore, for this category of surfaces with deformations close to isometry, the corresponding mappings between their conformal images are close to identity.

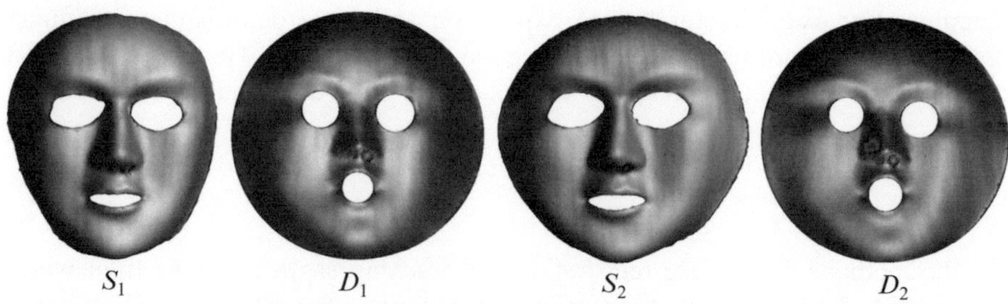

Fig. 5.1 Conformal mappings of two nearly isometric surfaces

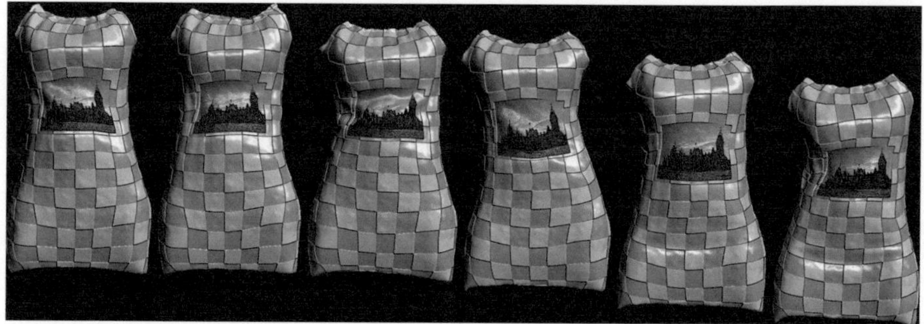

Fig. 5.2 Tracking for deformable clothes sequences. The consistent tracking results can be observed from the checkerboard texture motion

Application

Deformable Clothes Tracking. The clothes deformation is close to isometry. The deformable clothes tracking can be performed based on the registrations frame by frame. The clothes surfaces are captured by the 3D scanner introduced by [6]. Each frame is a topological quadrilateral (i.e., only one simple boundary with four corners), and can be mapped to a rectangle automatically by the Euclidean discrete surface Ricci flow. In order to verify if the deformation is close to be conformal/isometric, we compute the conformal module of each frame. The conformal modules are very consistent across the frames. This validates our hypothesis that the deformations are close to be conformal.

We also use vision techniques to find textural features on each frame and use them as hard constraints for the registrations between two adjacent frames. The registration between two planar rectangles with feature point constraints is performed based on planar harmonic maps. Because the deformations are close to be conformal, the harmonic mappings are close to identity. In order to visualize the tracking results, we put checkerboard textures to the first frame and propagate the texture parameters to the other frames through the correspondences from tracking. As shown in Fig. 5.2, the checkerboard textures are consistent across frames, without oscillating effects or checker collapse. This demonstrates that the registration between two frames is a diffeomorphism. The tracking is stable and automatic.

5.2.2 *Harmonic Mapping*

Harmonic mappings are commonly applied for computing diffeomorphisms between surfaces. There are three key factors to ensure a harmonic mapping to be diffeomorphic: (a) the convexity of the target domain, (b) the boundary condition, and (c) the Riemannian metric on the target surface. In the following, we discuss various harmonic mapping algorithms for surfaces with different topologies and metrics, which guarantee the output mappings to be diffeomorphisms.

Topological Disks

Suppose S_1 and S_2 are topological disks. We first map S_2 to the unit disk or other convex planar domain D using Ricci flow. The problem is converted to computing a harmonic mapping from the surface to the convex planar domain, $f : S_1 \rightarrow D$.

First, we construct a homeomorphism which maps the boundary of the source to the boundary of the target disk, $h : \partial S_1 \rightarrow \partial D$. For example, we denote the boundary of S_1 as $\gamma(s)$, where s is arc length parameter, and construct the boundary homeomorphism

$$h(\gamma(s)) = (\cos\theta(s), \sin\theta(s)), \ \theta(s) = 2\pi\frac{s}{c}, \tag{5.7}$$

where c is the total length of ∂S_1.

Second, we solve Laplace equation to get the harmonic mapping, $f(p) = (u(p), v(p))$, where u, v are harmonic functions on the surface with the Dirichlet boundary condition

$$\begin{cases} \Delta_g(u(p), v(p)) = (0,0) \ \ \forall p \notin \partial S_1 \\ (u(p), v(p)) \quad = h(p) \ \ \forall p \in \partial S_1 \end{cases}, \tag{5.8}$$

where $\Delta_{\mathbf{g}}$ is the Laplace–Beltrami operator induced by the Riemannian metric $\mathbf{g}$ on S_1. Choose an isothermal coordinates system z, $\mathbf{g} = e^{2\lambda}dzd\bar{z}$, then the *Laplace–Beltrami operator* is represented as

$$\Delta_{\mathbf{g}} = \frac{4}{e^{2\lambda}}\frac{\partial^2}{\partial z\partial\bar{z}} = \frac{1}{e^{2\lambda}}\left(\frac{\partial^2}{\partial x^2} + \frac{\partial^2}{\partial y^2}\right). \tag{5.9}$$

Since conformal metric deformations do not affect harmonic mappings, we can change the metric on the source $e^{2\lambda}dzd\bar{z}$ to $dzd\bar{z}$, then the above (5.8) becomes a classical Laplace equation on the plane.

In the discrete case, we approximate S_1 by a triangular mesh M. Suppose $[v_i, v_j, v_k]$ is a face on the mesh and the corner angle at vertex v_i is θ_i^{jk}. Suppose $[v_i, v_j]$ is an interior edge adjacent to two faces $[v_i, v_j, v_k]$ and $[v_j, v_i, v_l]$. The *cotangent weight* of the edge is given by

$$w_{ij} := \cot\theta_k^{ij} + \cot\theta_l^{ji}. \tag{5.10}$$

Algorithm 5.5 Harmonic map for topological disks

Input: A triangular mesh M which is a topological disk.
Output: A discrete harmonic mapping from M to the unit disk, $f : M \to \mathbb{D}$.
 1: Trace the boundary of the mesh, sort the boundary vertices $\{v_0, v_1, \ldots, v_{n-1}, v_n\}$, $v_n = v_0$, along the boundary.
 2: Compute the total length of the boundary

$$c = \sum_{k=0}^{n-1} len([v_k, v_{k+1}]).$$

 3: **for all** boundary vertex $v_k \in \partial M$, $k = 1..n$ **do**
 4:     Set $f(v_k) = (\cos \theta_k, \sin \theta_k)$, where $\theta_k = \frac{2\pi}{c} \sum_{i=0}^{k-1} len([v_i, v_{i+1}])$.
 5: **end for**
 6: **for all** edge $[v_i, v_j] \in M$ **do**
 7:     Compute the edge weight w_{ij} using (5.10) and (5.11).
 8: **end for**
 9: **for all** interior vertex $v_i \notin \partial M$ **do**
10:     Construct the linear equation: the discrete Laplacian at v_i is zero using (5.12).
11: **end for**
12: Solve the linear system, obtain the discrete harmonic map.

If $[v_i, v_j]$ is a boundary edge adjacent to one face $[v_i, v_j, v_k]$, then the *cotangent weight* is given by

$$w_{ij} := \cot \theta_k^{ij}. \tag{5.11}$$

Let $f : V \to \mathbb{R}$ be a function defined on the vertices of the mesh. The *discrete Laplace–Beltrami operator* is defined as

$$\Delta f(v_i) = \sum_{[v_i, v_j] \in M} w_{ij} (f(v_j) - f(v_i)). \tag{5.12}$$

We say f is a *discrete harmonic function* if $\Delta f(v_i) = 0$ for all interior vertex $v_i \notin \partial M$. Algorithm 5.5 explains the details for computing the harmonic map for a topological disk.

Topological Spheres

Suppose both the source and target surfaces are topological spheres. By using Ricci flow, we deform the target surface to a unit sphere. The mapping from a surface S to a unit sphere, $f : S \to \mathbb{S}^2$, can be decomposed to three coordinate functions, $f = (f_1, f_2, f_3)$. The Laplacian of f is a vector valued function, $\Delta_{\mathbf{g}} f = (\Delta_{\mathbf{g}} f_1, \Delta_{\mathbf{g}} f_2, \Delta_{\mathbf{g}} f_3)$. Then f is a harmonic map if and only if $\Delta_{\mathbf{g}} f(p)$, $p \in M$ is along the normal at $f(p) \in \mathbb{S}^2$ on the target, which is $f(p)$ itself,

$$\Delta_{\mathbf{g}} f - \langle \Delta_{\mathbf{g}} f, f(p) \rangle f(p) = 0.$$

The spherical harmonic map can be computed as follows. First, the initial map is constructed as the *Gauss map* from S to the unit sphere. Then we use the *nonlinear heat diffusion* method to update the mapping,

$$\frac{\partial f(p,t)}{\partial t} = -[\Delta_{\mathbf{g}} f(p,t) - \langle \Delta_{\mathbf{g}} f(p,t), f(p,t)\rangle f(p,t)].$$

Since the harmonic mapping is not unique and all harmonic mappings differ by a spherical Möbius transformation, we add the following normalization condition to ensure the uniqueness of the solution,

$$\int_S f(p)dp = \mathbf{0}.$$

Algorithm 5.6 explains the details for computing the harmonic map for a topological sphere.

Topological Tori

Suppose $(S_1, \mathbf{g}_1)$ and $(S_2, \mathbf{g}_2)$ are two genus one closed surfaces with Riemannian metrics. Using Ricci flow, we can deform them to flat tori $\mathbb{E}^2/\Lambda_1$ and $\mathbb{E}^2/\Lambda_2$, where the lattices

$$\Lambda_1 = \{z \to z + m + n\omega_1, m, n \in \mathbb{Z}\}, \ \Lambda_2 = \{w \to w + m + n\omega_2, m, n \in \mathbb{Z}\}.$$

Then the harmonic mapping between two flat tori is an affine map,

$$w = \frac{\omega_2 - \bar{\omega}_1}{\omega_1 - \bar{\omega}_1}z + \frac{\omega_1 - \omega_2}{\omega_1 - \bar{\omega}_1}\bar{z}.$$

High Genus Closed Surfaces

Suppose $(S_1, \mathbf{g}_1)$ and $(S_2, \mathbf{g}_2)$ are two high genus closed surfaces with Riemannian metrics. Using Ricci flow, we can deform their metrics to be hyperbolic. First, we construct an initial map, then use the *hyperbolic heat flow* method to diffuse it to a harmonic map.

The initial diffeomorphism can be constructed using divide-and-conquer approach. Suppose the surface is of genus g. Then it can be decomposed to $2g - 2$ pairs of pants. A pair of pants is a genus zero surface with three boundaries. This is called a *pants decomposition* of the surface. The pants decomposition process is straightforward. Compute a cut graph C of the current surface using Algorithm 5.3. If C includes a loop γ which is not homotopic to any boundary loop, then slice the surface along γ. If the surface is still connected, then repeat this procedure; if the

---

Algorithm 5.6 Harmonic map for topological spheres

Input: A triangular mesh M which is a topological sphere, the curvature error threshold ϵ, and the step length λ.

Output: A discrete harmonic mapping from M to the unit sphere, $f : M \to \mathbb{S}^2$.

1: Initialize f as the Gauss map.
2: Compute the edge weight w_{ij} using (5.10) and (5.11).
3: **for all** vertex $v_i \in M$ **do**
4:     Compute the weight for v_i

$$w_i = \frac{1}{3} \sum_{[v_i,v_j,v_k]\in M} A([v_i,v_j,v_k]),$$

    where $A([v_i,v_j,v_k])$ is the area of the face $[v_i,v_j,v_k]$.
5: **end for**
6: **while** $\max_{v_i \in M} |h(v_i)| > \epsilon$ **do**
7:     **for all** vertex $v_i \in M$ **do**
8:         Compute the discrete Laplacian $\Delta f(v_i)$ using (5.9).
9:         Project the Laplacian to the tangent plane at $f(v_i)$,

$$h(v_i) = \Delta f(v_i) - \langle \Delta f(v_i), f(v_i) \rangle f(v_i).$$

10:         Update the image of v_i

$$f(v_i) = f(v_i) - \lambda h(v_i).$$

11:     **end for**
12:     Compute the mass center

$$c = \sum_i f(v_i)w_i / \sum_i w_i.$$

13:     **for all** vertex $v_i \in M$ **do**
14:         Update the image of $f(v_i)$

$$f(v_i) = (f(v_i) - c)/|f(v_i) - c|.$$

15:     **end for**
16: **end while**

---

surface is split into two connected components, then repeat this procedure on each of them. Until all the loops are homotopic to one of the boundary loops, then each connected component is a pair of pants. Suppose the surface is with a hyperbolic metric, so that each homotopy class has a unique geodesic loop. All the cut loops can be chosen to be geodesics. Eventually, the surface is decomposed to $2g - 2$ pairs of hyperbolic pants, as shown in Fig. 5.3**a**.

Given a pair of hyperbolic pants with three geodesic boundaries $\{\gamma_i, \gamma_j, \gamma_k\}$, it can be further divided into two identical hyperbolic hexagons. As shown in Fig. 5.3 **b**, let τ_i be the unique geodesic orthogonal to both γ_j and γ_k, then τ_i is the shortest path between these two boundary loops. Similarly, τ_j, τ_k are the shortest paths between $\{\gamma_k, \gamma_i\}$ and $\{\gamma_i, \gamma_j\}$, respectively. Then $\{\tau_i, \tau_j, \tau_k\}$ separate the surface

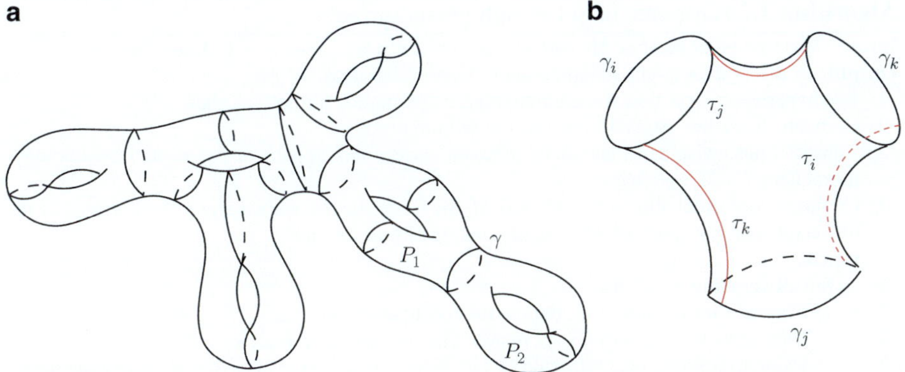

Fig. 5.3 Pants decomposition of a high genus surface (**a**) and a pair of hyperbolic pants (**b**)

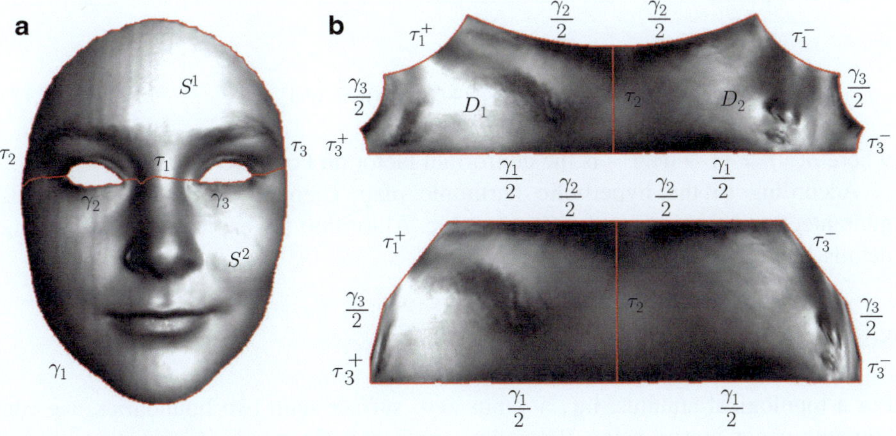

Fig. 5.4 Pants decomposition and hyperbolic embedding. By hyperbolic Ricci flow, a triply connected domain (**a**) is mapped to two hyperbolic hexagons, embedded on the Poincaré disk and Klein disk, respectively (**b**)

into two hyperbolic geodesic hexagons. If we use Klein model for hyperbolic plane $\mathbb{H}^2$, then all the hyperbolic hexagons can be represented as convex Euclidean planar hexagons, as shown in Fig. 5.4.

We can perform consistent hyperbolic pants decompositions for both the source and the target surfaces, decompose each pair of pants to two hyperbolic hexagons, then match the corresponding hexagons on Klein model using harmonic maps on Euclidean plane with consistent boundary conditions. Now we obtain an initial homeomorphism. Then we use the *hyperbolic heat diffusion* method to deform the initial mapping to be harmonic. Under the hyperbolic metric, isometrically embed the universal covering spaces of both the source and the target to the Poincaré disk. This gives isothermal coordinates z for the source S_1 and w for the target S_2. The hyperbolic heat diffusion is given by

---

Algorithm 5.7 Harmonic map for high genus surfaces

Input: Two triangular meshes M_1 and M_2, both of them are genus $g > 1$ closed meshes.
Output: A discrete harmonic mapping under hyperbolic metric of M_2.
  1: Use hyperbolic Ricci flow to compute hyperbolic metrics for both meshes.
  2: Compute consistent pants decomposition of both meshes.
  3: Construct piecewise harmonic maps between corresponding pairs of pants, get initial homeomorphism, $f : M_1 \rightarrow M_2$.
  4: Compute conformal atlases for M_1 and M_2 by isometrically embedding finite portions of the universal covering space of M_1 and M_2 onto the Poincaré disk.
  5: **repeat**
  6:　　**for all** vertex $v_i \in M_1$ **do**
  7:　　　　Choose a local chart of v_i, denote the local parameter as z,
  8:　　　　Choose a local chart of $f(v_i)$, denote the local parameters as w,
  9:　　　　Using hyperbolic heat diffusion (5.13) $\frac{\partial w}{\partial t} = -(w_{z\bar{z}} + (\log \rho)_w w_z w_{\bar{z}})$ to update the image of v_i.
 10:　　**end for**
 11: **until** $|w_{z\bar{z}} + (\log \rho)_w w_z w_{\bar{z}}| < \epsilon$.

$$\frac{\partial w}{\partial t} = -(w_{z\bar{z}} + (\log \rho)_w w_z w_{\bar{z}}), \tag{5.13}$$

where $\rho(w) = (1 - w\bar{w})^{-2}$ is the conformal factor on Poincaré disk.

According to the hyperbolic harmonic map Theorem 3.10, the solution is guaranteed to be unique and diffeomorphic. Algorithm 5.7 explains the computing details.

Surfaces with Boundaries

For a topological annulus, i.e., a genus zero surface with two boundaries, we can compute a flat metric using Ricci flow, such that the geodesic curvatures along the boundaries are zero everywhere. Then the annulus can be represented as $\{z \in \mathbb{C} | Img(z) \leq 1\}/\Lambda$, where the lattice transformation group $\Lambda = \{z | z \rightarrow z + m\omega\}$, $\omega \in \mathbb{R}$. Therefore, the fundamental domain is a rectangle. The harmonic mapping between two annuli can be given by an affine mapping directly.

For a topological multiply connected annulus, i.e., a genus zero surface with more than two boundaries, we can compute the hyperbolic metric using Ricci flow, such that all the boundaries become geodesics. Then we decompose it to pairs of pants, isometrically embed the pairs of pants in Klein disk and use piecewise Euclidean harmonic maps to construct the initial map. Finally, we can use hyperbolic heat diffusion as (5.13) to get the global hyperbolic harmonic map. Figure 5.4 shows the embedding of a triply connected domain (a pair of pants) in both Poincaré disk and Klein disk. In Klein disk, hyperbolic hexagons are identical to Euclidean convex hexagons.

For high genus surfaces with boundaries, we can use the method similar to that for topological multiply connected annuli, or use double covering technique explained in Sect. 4.1.3 to convert the surface to a symmetric closed surface and then apply the hyperbolic harmonic map technique for the doubled surface.

Application

Human Face Registration with Expressions. In general, human facial expression changes are not conformal, and far away from isometry. This can be easily verified by computing the conformal modules of the same face with different expressions. The expression surfaces can be obtained from a real-time structure light 3D scanner [17]. For example, the face surfaces without eyes and mouth interior regions are conformally mapped to the planar circle domains, then by comparing the inner circle positions and sizes, one can tell the change of the conformal structures [28]. It is challenging to register the circle domains using harmonic maps, because the harmonic maps may not be diffeomorphic due to their concavities. Instead, we can apply hyperbolic Ricci flow and use hyperbolic harmonic map to ensure the diffeomorphic results. We can also use Euclidean Ricci flow to concentrate all the curvatures to some feature points, and set the Gauss curvature and geodesic curvature to be zeros everywhere, then the surface can be decomposed to the union of planar convex polygons and registered by piecewise harmonic maps. Details can be found in [28]. Figure 5.5 demonstrates the registration for a sequence of human facial surfaces with non-isometric deformations, complex topologies, and inconsistent boundaries. The one-to-one correspondence is visualized by the consistent checkerboard texture mapping results.

Beating Heart Tracking. Dynamic shape tracking is very important in morphology analysis, abnormality detection and prediction, and real-time monitoring and guidance during radiotherapy and surgery. We consider a sequence of frames of the beating 3D heart surface from end diastole to end systole and aim to track the motion of the left ventricle part. The 3D geometry of heart surfaces can be reconstructed from the tagged MRI data [4]. We apply the Euclidean Ricci flow algorithm to map each heart surface frame to its canonical planar domain and register each two adjacent frames by mapping the corresponding planar domains. Although the nonrigid deformation of the heart is significant between different frames, the Ricci flow method still captures the deformation almost indistinguishably from the ground truth. Details can be found in [28].

5.2.3 Quasi-Conformal Mapping

Given the source and target surfaces $(S_k, \mathbf{g}_k)$ with the same topologies, fix the homotopy class. All the diffeomorphisms form the mapping space, which is essentially equivalent to the functional space of the Beltrami coefficients (or Beltrami differentials). To simplify the discussion, we assume both the source and the target are topological disks, but the method is general to surfaces with arbitrary topologies.

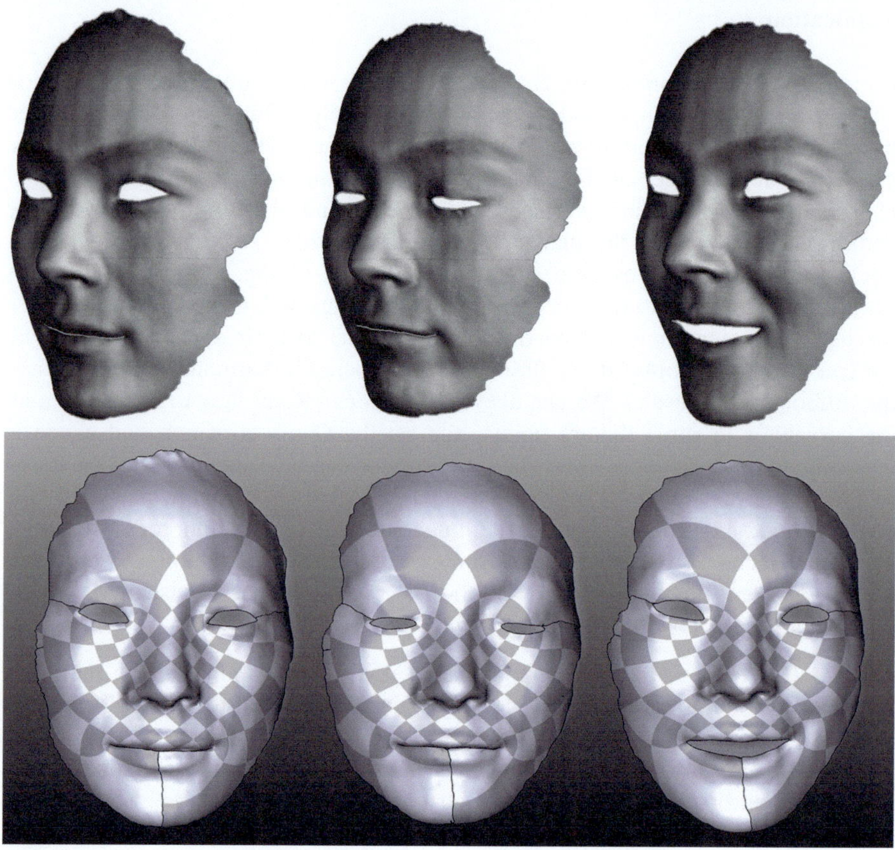

Fig. 5.5 Registration for human facial surfaces with expression change by hyperbolic Ricci flow

Solving Beltrami Equation

We use Ricci flow to conformally map the source and the target to the unit disk, $\phi_k : S_k \to \mathbb{D}$, then construct an automorphism for the disk itself: $f : \mathbb{D} \to \mathbb{D}$, $w = f(z)$. The mapping space is equivalent to the Beltrami coefficient space,

$$\frac{\{\text{Diffeomorphisms}\}}{\{\text{Mobius}\}} \cong \{\mu : \mathbb{D} \to \mathbb{C} | \|\mu\|_\infty < 1\}.$$

Assume the Beltrami coefficient μ is given. The mapping can be obtained by using the *auxiliary metric method* as explained in Sect. 3.4.1. Basically, we construct a new Riemannian metric $|dz + \mu d\bar{z}|^2$ and compute a conformal mapping $f : (\mathbb{D}, |dz + \mu d\bar{z}|^2) \to (\mathbb{D}, |dw|^2)$ using Ricci flow with normalization condition to remove Möbius ambiguity, then f satisfies the Beltrami equation $f_{\bar{z}} = \mu f_z$.

Algorithm 5.8 illustrates the pipeline. The details for computing the general diffeomorphisms using the auxiliary metric method can be found in [25, 26].

Algorithm 5.8 Auxiliary metric method for solving Beltrami equation

---

Input: A triangular mesh M, which is topological disk; Beltrami coefficient $\mu : V \to \mathbb{C}$.
Output: Quasi-conformal mapping $f : M \to \mathbb{D}$ satisfying Beltrami equation with μ.
 1: Compute a conformal mapping $\phi : M \to \mathbb{D}$ using Ricci flow.
 2: **for all** edge $[v_i, v_j] \in M$ **do**
 3:     Compute the edge length l_{ij} using the induced Euclidean metric.
 4:     Compute the derivative of conformal coordinates on $[v_i, v_j]$, $dz_{ij} \leftarrow \phi(v_j) - \phi(v_i)$.
 5:     Compute the Beltrami coefficient on $[v_i, v_j]$, $\mu_{ij} \leftarrow \frac{1}{2}(\mu(v_i) + \mu(v_j))$.
 6:     Compute the scalar of metric $\lambda_{ij} \leftarrow \frac{|dz_{ij} + \mu_{ij} d\bar{z}_{ij}|}{|dz_{ij}|}$.
 7:     Compute the auxiliary metric $\tilde{l}_{ij} \leftarrow \lambda_{ij} l_{ij}$.
 8: **end for**
 9: Compute the Riemann mapping f using Ricci flow with normalization condition, based on the auxiliary metric.

Optimization in Mapping Space

In practice, it is always desirable to find an optimal diffeomorphism minimizing some specific energy forms. This requires the optimization techniques in the space of all the diffeomorphisms between the source and the target within a homotopy class, namely, the mapping space, or equivalently, the Beltrami coefficient space. *Beltrami holomorphic flow* is one of the practical optimization methods in the mapping space. In the following, we assume that both surfaces are topological spheres and uniformized to the extended complex plane $\bar{\mathbb{C}}$ by Ricci flow, and $f^\mu : S_1 \to S_2$ is a mapping associated with the Beltrami coefficient μ.

Suppose the variation of the Beltrami coefficient is

$$\tilde{\mu}(z) = \mu(z) + t\nu(z) + O(t^2), \ z = x + iy.$$

According to quasi-conformal geometry theory, the variation of the mapping is

$$f^{\tilde{\mu}}(z) = f^\mu(z) + tV(f^\mu, \nu)(z) + O(t^2),$$

where the deformation field

$$V(f^\mu, \nu)(z) = \int_{\bar{\mathbb{C}}} K(\zeta, z) d\zeta_x d\zeta_y,$$

where $\zeta = \zeta_x + i\zeta_y$, and

$$K(\zeta, z) = -\frac{f^\mu(z)(f^\mu(z) - 1)}{\pi} \left(\frac{\nu(\zeta)((f^\mu)_z(\zeta))^2}{f^\mu(\zeta)(f^\mu(\zeta) - 1)(f^\mu(\zeta) - f^\mu(z))} \right).$$

This gives an explicit way to perform optimization in the mapping space. In the following, we give a simple example for illustration. Let $h^k : S_k \to \mathbb{R}$ be some functions on both the source and the target. We define a simple energy form

$$E_h(\mu) := \int_{\mathbb{C}} (h^1(z) - h^2 \circ f^\mu(z))^2 dx dy,$$

and use $w = f^\mu(z)$ to represent the mapping, then

$$\frac{d}{dt}E_h(\mu + tv) = -2\int_{\mathbb{C}}(h^1(z) - h^2(w(z)))\left(h_w^2\frac{dw}{dt} + h_{\bar{w}}^2\frac{d\bar{w}}{dt}\right)dxdy,$$

where

$$\frac{dw(z)}{dt} = -\frac{w(z)(w(z)-1)}{\pi}\int_{\mathbb{C}}\frac{w_z^2(\zeta)v(\zeta)}{w(\zeta)(w(\zeta)-1)(w(\zeta)-w(z))}d\zeta_xd\zeta_y.$$

Define an auxiliary function $F(\mu,\zeta)$ as

$$F(\mu,\zeta) = \frac{w_z(\zeta)^2}{w(\zeta)(w(\zeta)-1))}\int_{\mathbb{C}}\frac{(h^1(z)-h^2(w(z)))h_w^2(w(z))}{\pi(w(z)-w(\zeta))}w(z)(w(z)-1)dxdy,$$

then

$$\frac{d}{dt}E_h(\mu + tv) = -2\int_{\mathbb{C}}F(\mu,\zeta)v(\zeta) + \overline{F(\mu,\zeta)}\bar{v}(\zeta)d\zeta_xd\zeta_y.$$

Therefore, the flow

$$\frac{d}{dt}\mu(\zeta) = -\overline{F(\mu,\zeta)}$$

will minimize the energy $E_h(\mu)$. This gives an optimization method in the mapping space.

In practice, the energy can be specially designed for different purposes. If we want to get a mapping as isometric as possible, then we can set the energy to be

$$E_\lambda(\mu) := \int(\lambda_1 - \lambda_2 \circ f^\mu)^2,$$

where λ_k are the conformal factors. If $E_\lambda(\mu)$ is zero, then f^μ is an isometric mapping. If we prefer the mapping to be a rigid motion in $\mathbb{E}^3$, then we set $E(\mu) = E_\lambda + E_H$, where

$$E_H(\mu) := \int(H_1 - H_2 \circ f^\mu)^2,$$

where H_k are the mean curvature functions on the source and the target surfaces. According to Theorem 3.3, $E(\mu)$ is zero if and only if the source and the target surfaces differ by a rigid motion in $\mathbb{E}^3$. In order to get higher continuity of the mapping, we can set the energy to be the L^p norm of the Beltrami coefficient. The minimizer for L^∞ norm is the Teichmüller map. If we know the corresponding feature points $\{z_i\}$ and $\{w_i\}$ on the source and the target, respectively, and would like the mapping to match them, then we can add another term $\sum_i |w_i - f^\mu(z_i)|^2$.

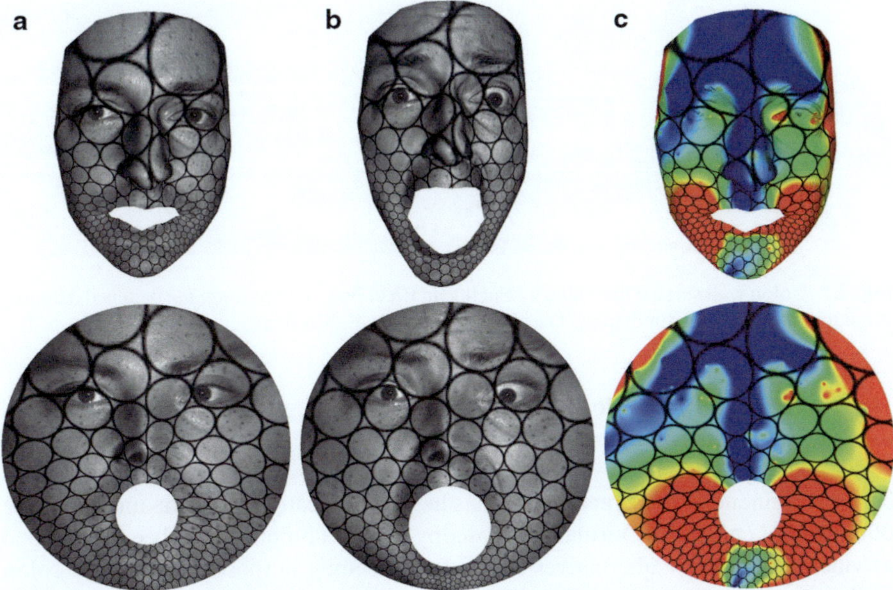

Fig. 5.6 Quasi-conformal registration for a topological annulus case. The ellipse field on the source domain (**a**) is mapped to the circle field on the target domain (**b**). The distortion is visualized by the eccentricity of ellipses $D(\mu)$ on both the 3D surface and its circle domain (**c**), where the *red* color denotes the higher non-conformality and the *blue* color denotes the higher conformality

Application

Registration of Surfaces with Large Deformations. Quasi-conformal geometry provides a powerful approach to registration. The strategy for registration is first to estimate the Beltrami coefficient μ and then compute the quasi-conformal mapping under the auxiliary metric associated with μ. A coarse-to-fine method to compute μ was introduced in [21]. The Beltrami coefficient is estimated from the graph of feature points, then refined by the diffusion on the whole surface. The Beltrami coefficient can also be obtained by minimizing the matching energy [13]. The methods can handle surfaces with complicated topologies and large anisometric deformations.

Figure 5.6 shows the registration between two facial surfaces of the same person with a large expression deformation. Both faces are mapped to the planar annuli conformally, then the annuli are registered by a quasi-conformal mapping using the auxiliary metric method. The consistent circle packing texture mapping demonstrates the one-to-one correspondence between two faces. From the figure, we can see that an ellipse field on the calm face is mapped to a circle field on the frightened face, therefore, the mapping is quasi-conformal. The distortion can be visualized by the color encoded eccentricity and argument of the Beltrami coefficient.

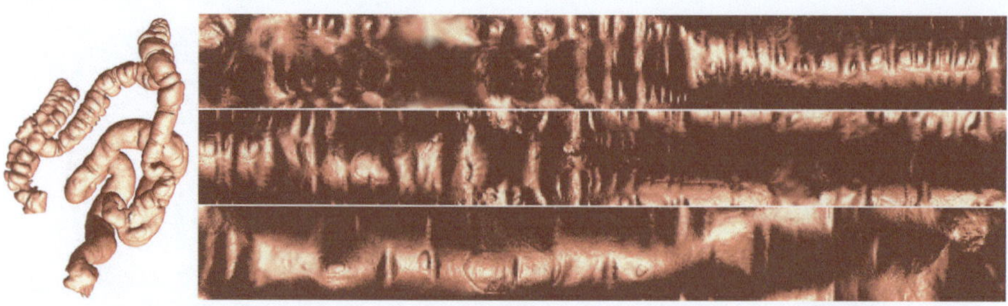

Fig. 5.7 Colon wall surface flattening by Euclidean Ricci flow. The *rectangular* flattened mapping is cut to three segments for display. Data courtesy of Arie E. Kaufman

Virtual Colonoscopy. Colorectal cancer is the third most incident cancer and the second leading cause of cancer-related mortality in the USA [8]. Optical colonoscopy (OC), whereby precancerous polyps can be located and removed, has been recommended for screening and has helped to greatly reduce the mortality rate of colorectal cancer [1]. Virtual colonoscopy (VC) techniques have been developed as viable noninvasive alternatives to OC for screening purposes [7, 11]. The colon wall surface is reconstructed from CT images. By using Ricci flow, one can flatten the whole colon wall surface onto a planar rectangle [15]. Then polyps and other abnormalities can be found efficiently on the planar image by Computer Aided Detection (CAD) techniques.

For a VC procedure, CT scans of the abdomen are commonly acquired with the patient in both supine (facing up) and prone (facing down) positions. Being able to register these two scans is useful for a routine VC system to provide the user with ability for confirming a finding or easy comparison when something might be unclear in one of the scans, and is also helpful for a CAD system to achieve greater accuracy while at the same time reducing false positive results. As shown in Fig. 5.7, the colon surface is regarded as a topological cylinder, therefore can be mapped to a rectangle using Euclidean Ricci flow. By quasi-conformal registration, we obtain the one-to-one correspondence between them, then use that to analyze and visualize the elastic deformation of the colon. Details can be found in [27].

Human Facial Expression Deformation. Different human facial expressions have different deformation patterns, therefore, different types of Beltrami coefficients. We can do clustering in the Beltrami coefficient space for human facial expression analysis and recognition [21].

Brain Morphology. Neuroscientists commonly aim to identify abnormal deformations of cortical and subcortical structures in the brain in order to detect systematic patterns of alterations in brain diseases. The normal deformations of a brain cortical surface are isotropic, namely, nearly conformal. Some diseases will cause anisotropic deformations at some regions. Then for areas with isotropic deformations, the norm of Beltrami coefficient μ is close to zero; for areas with anisotropic deformation, μ has greater norms. For example, the norm of μ is a good indicator for the gyral thickening on the cortical surface.

This method can be generalized to monitor the progression of neuron diseases by qualitatively measuring the degree of abnormalities using the time-dependent Beltrami coefficients $\mu(t)$. We reconstruct a sequence of cortical surfaces $S(t)$ of the same patient from regular MRI scans, and register $S(t)$ to the initial surface $S(0)$. The Beltrami coefficient of the mapping is denoted as $\mu(t)$. By analyzing $\mu(t)$, we can monitor the shape deformations and estimate the severity of abnormalities. Details can be found in [12].

5.3 Shape Analysis

The purpose of shape analysis is to classify shapes, compare shapes, quantify their similarities and differences, and measure the distances between shapes. One systematic way to perform shape analysis is to construct the *shape space*, which is an infinite dimensional Riemannian manifold. Each point in the shape space represents a shape, or a class of shapes; a curve represents a deformation process from one shape to the other. The Riemannian metric is used to measure the distances and compute geodesics, where geodesics represent the most natural deformation from the source to the target.

In the following discussion, we focus on three models of shape space:

1. *2D planar shape space based on conformal welding theory*: The planar contours are converted to the diffeomorphism space of the unit circle, and the distance between shapes is measured using the Weil–Petersson metric. The basic idea is straightforward. A planar contour segments the whole plane to two connected components. Each of them is mapped to the unit disk via Riemann mapping. This induces a homeomorphism between the boundaries of the two disks, namely, an automorphism of the unit circle, which is called the "fingerprint" of the contour. Inversely, we can weld two disks using the fingerprint to fully recover the original contour. We can measure the distance between two contours by computing the norm between their fingerprints. Details can be found in Sect. 5.3.1.

2. *Teichmüller space for surfaces*: All surfaces are classified by conformal equivalence relation. The coordinates of a shape is given by its conformal module. The distance is measured by Teichmüller metric. The conformal modules can be computed using the Ricci flow method. The Teichmüller distance requires solving the extremal quasi-conformal mapping problem. The method is elaborated in Sect. 5.3.2.

3. *3D surface space based on conformal representation*: For surfaces embedded in $\mathbb{R}^3$, they are determined by the conformal factor, the mean curvature, and the boundary conditions, uniquely up to a rigid motion. Then we can represent a surface by its conformal factor and mean curvature functions on the uniformization domain. This unified view greatly simplifies the task of shape analysis. The method is explained thoroughly in Sect. 5.3.3.

5.3.1 2D Shape Space Based on Conformal Welding

Ricci flow can be applied for modeling the 2D shape space based on *conformal welding* theory. Suppose $\Gamma = \{\gamma_0, \gamma_1, \ldots, \gamma_n\}$ is a set of nonintersecting smooth closed curves on the complex plane. Γ segments the plane to a set of connected components $\{\Omega_0, \Omega_1, \ldots, \Omega_{n+1}\}$. Each segment Ω_k may be a simply connected domain or a multiply connected domain.

We assume Ω_{n+1} contains the infinity point. Select one point p outside Ω_{n+1}, $p \notin \Omega_{n+1}$. By using a Möbius transformation, $\phi(z) = 1/(z-p)$, p is mapped to ∞, and Ω_{n+1} is mapped to a compact domain. Replace Ω_{n+1} by $\phi(\Omega_{n+1})$. Construct conformal mappings using Ricci flow, $\phi_k : \Omega_k \to D_k$, to map each segment Ω_k to a circle domain D_k, $0 \le k \le n + 1$. We denote the conformal module of D_k as $\mathrm{Mod}(D_k)$, which is also the conformal module of Ω_k, $\mathrm{Mod}(\Omega_k)$. Assume $\gamma_k \in \Gamma = \Omega_i \cap \Omega_j$. Then $\phi_i(\gamma_k)$ is a circular boundary C_k^i of the circle domain D_i, and $\phi_j(\gamma_k)$ is also a circular boundary C_k^j of D_j. Let $f_k := \phi_i \circ \phi_j^{-1}|_{\mathbb{S}^1} : \mathbb{S}^1 \to \mathbb{S}^1$ be the diffeomorphism from the circle $\mathbb{S}^1$ to itself, which is called the *signature of γ_k*.

Definition 5.1 (Conformal Welding Signature of a Family of Planar Curves). The signature of a family of nonintersecting closed planar curves $\Gamma = \{\gamma_0, \gamma_1, \ldots, \gamma_n\}$ is defined as $S(\Gamma) := \{f_0, f_1, \ldots, f_n\} \cup \{\mathrm{Mod}(D_0), \mathrm{Mod}(D_1), \ldots, \mathrm{Mod}(D_{n+1})\}$.

Theorem 5.1 (Conformal Welding of Planar Closed Curves [14]). *The family of smooth planar closed curves Γ is determined by its signature $S(\Gamma)$, uniquely up to a Möbius transformation of the Riemann sphere $\mathbb{C} \cup \{\infty\}$.*

Proof. Let Ω_k be one segment, which is conformally mapped to a circle domain D_k. Assume two segments Ω_i and Ω_j share a common boundary γ_k, $\gamma_k = \Omega_i \cap \Omega_j$. We can glue D_i and D_j using the signature of γ_k. Namely, a boundary point $p \in \partial\Omega_i$ is equivalent to $q \in \partial\Omega_j$, $p \sim q$, if and only if $q = f_k(p)$. Then we glue all the circle domains to get a Riemann surface

$$R := \bigcup_k D_k / \sim .$$

R is a topological sphere with a conformal structure. Then by uniformization theorem, it can be equipped with the spherical metric. The unit sphere is conformally mapped to the extended complex plane $\bar{\mathbb{C}}$. By a Möbius transformation, the images of the boundaries of D_k's coincide with γ_k's. $\qquad\square$

The theorem states that the conformal welding signature determines 2D shapes uniquely up to a Möbius transformation. The shape signature $S(\Gamma)$ gives us a complete representation for the 2D shape. Therefore the 2D shape space can be modeled as the space of all conformal welding signatures. Furthermore, we can also define a Riemannian metric for the 2D shape space, such as the Weil–Petersson metric [16].

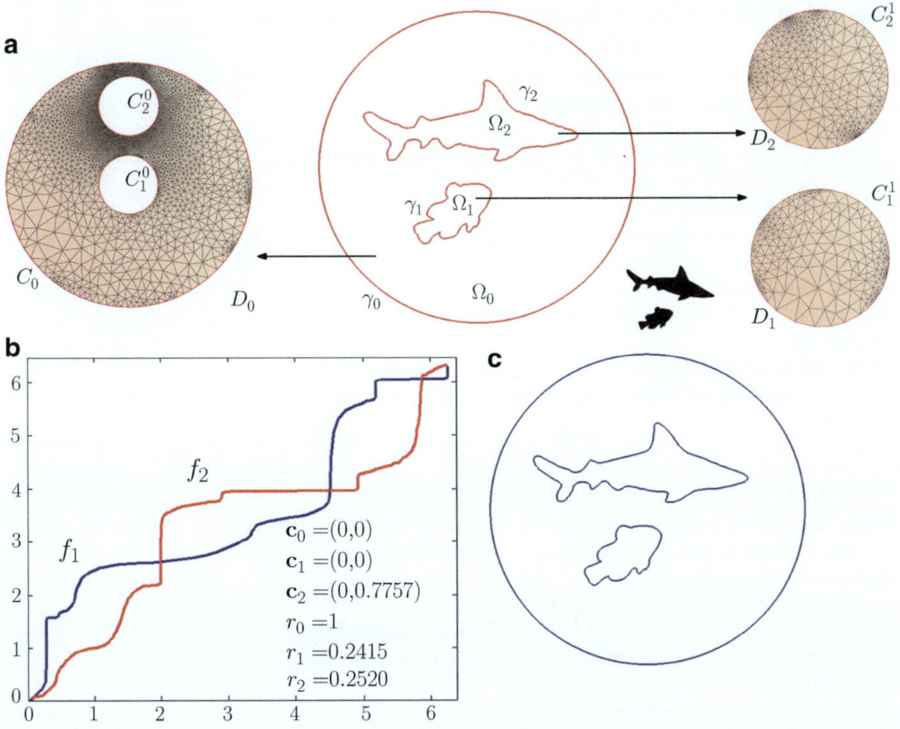

Fig. 5.8 Conformal welding signature. (**a**) Each segment is mapped to a circle domain. (**b**) The conformal modules (centers and radii of inner circles) of the circle domains and the diffeomorphisms of the circles define the signature. (**c**) The shape can be reconstructed from the signature

We demonstrate the process for computing the conformal welding signature with a double fish image as shown in Fig. 5.8. Given the original image, we first perform image segmentation, then calculate the contours of the segments in the image. For simplicity, we treat the outermost boundary of the image as the unit circle. Then all the contours segment the image to planar domains $\Omega_0, \Omega_1, \Omega_2$. We conformally map each planar domain to a circle domain, $\Phi_i : \Omega_i \rightarrow D_i$. Ω_0 is mapped to a disk D_0 with two circular holes. The centers and radii $(\mathbf{c}_0, r_0)$ and $(\mathbf{c}_1, r_1)$ form the conformal module of Ω_0. Also, Ω_1 and Ω_2 are mapped to the unit disks D_1 and D_2, respectively. The contour γ_1 of the small fish is mapped to the boundary C_1^1 of D_1 and one inner boundary C_1^0 of D_0. The signature is given by $f_1 := \Phi_1 \circ \Phi_0^{-1}$. Similarly, the signature f_2 of the contour of the shark can also be computed. The signature of both contours is given by $S(\Gamma) = \{\mathbf{c}_0, \mathbf{c}_1, r_1, r_2, f_1, f_2\}$. The computational steps are shown in Algorithm 5.9. The planar shapes can be reconstructed from the signature through the combinatorial operations as described in Algorithm 5.10.

Algorithm 5.9 Conformal welding signature of planar shapes

Input: A planar triangular mesh Ω with one exterior boundary γ_0 and n inner contours $\{\gamma_k, k = 1, \ldots, n\}$, which decompose Ω into $n + 1$ sub-domains $\{\Omega_k, k = 0, 1, \ldots, n\}$.
Output: Conformal welding signature

$$S(\Gamma) := \{f_1, \ldots, f_n\} \cup \{\mathrm{Mod}(\Omega_0), \mathrm{Mod}(\Omega_1), \ldots, \mathrm{Mod}(\Omega_n)\}.$$

1: **for all** sub-domain Ω_k **do**
2:     Compute the conformal map from Ω_k to circle domains D_k, $\phi_k : \Omega_k \to D_k$.
3: **end for**
4: **for all** sub-domain Ω_k **do**
5:     Compute the conformal module $\mathrm{Mod}(\Omega_k)$ of D_k.
6: **end for**
7: **for all** inner boundary $\gamma_k = \Omega_i \cap \Omega_j$ **do**
8:     Compute the signature of γ_k, $f_k = \phi_i \circ \phi_j^{-1}$.
9: **end for**

Algorithm 5.10 Planar shape reconstruction from conformal welding signature

---

Input: Conformal welding signature: the automorphisms of circles $\{f_1, \ldots, f_n\}$ and the conformal modules $\{\mathrm{Mod}(D_0), \mathrm{Mod}(D_1), \ldots, \mathrm{Mod}(D_n)\}$.
Output: Planar contours Γ.
1: **for all** conformal module $\mathrm{Mod}(D_k)$ **do**
2:     Construct circle domain D_k directly from its module $\mathrm{Mod}(D_k)$.
3:     Use Delaunay refinement algorithm to compute the meshing of D_k.
4: **end for**
5: **for all** circle automorphism f_k **do**
6:     Assume $\gamma_k = \Omega_i \cap \Omega_j$. Combinatorially glue the triangular mesh D_i and D_j by f_k. For each vertex $v_i \in \partial D_i$, insert $f_k(v_i)$ to ∂D_j; vice versa, for each vertex $v_j \in \partial D_j$, insert $f_k^{-1}(v_j)$ to ∂D_i.
7:     Use constrained Delaunay triangulation to refine the triangulation of D_i and D_j.
8:     Glue the refined triangle mesh D_i and D_j combinatorially by f_k.
9: **end for**
10: Run Ricci flow algorithm to compute a spherical metric for the glued mesh R.
11: Conformally map R to the complex plane using stereo-graphic projection. Normalize the mapping by a Möbius transformation.

Application

2D Shape Indexing and Retrieval. Shape analysis of objects from their observed silhouettes is important for many computer vision applications, such as classification, recognition, and image retrieval. In real world applications, objects from their observed silhouettes are usually multiply connected domains, i.e., domains with holes in the interior. Most of the existing methods work only on simple closed curves and generally cannot deal with multiply connected objects. Sharon and Mumford [16] proposed a conformal approach to model simple closed curves which captures subtle variability of shapes uniquely up to scalings and translations. They also introduced the Weil–Petersson metric on their proposed shape space. The method is generalized to represent multiple objects in a single image (i.e., with multiple

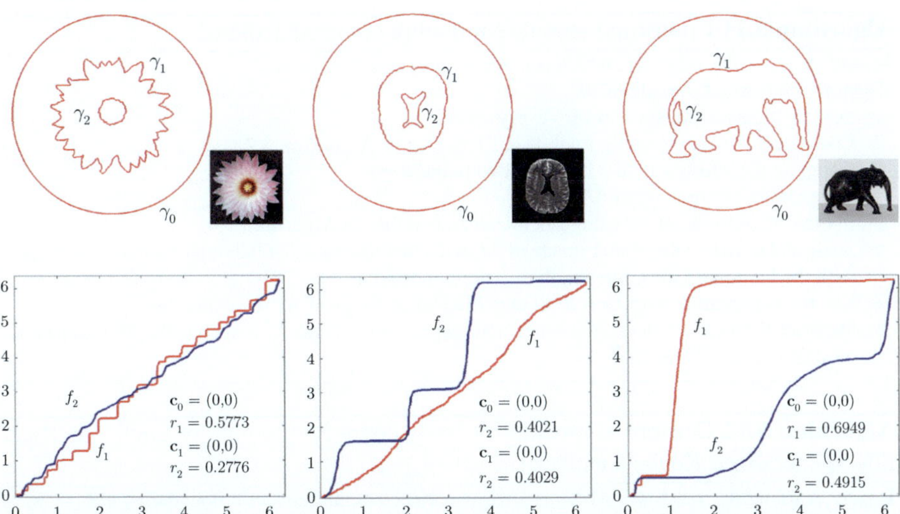

Fig. 5.9 Conformal welding shape signatures of three different images with two levels of contours

contours) based on conformal welding signatures in [14]. The shape signature is complete and can represent 2D shapes with arbitrary topologies uniquely up to scalings and translations. As shown in Fig. 5.9, the three different images with multiple boundaries have very different shape signatures. This demonstrates that the representation can be used to effectively classify 2D shapes.

Brain Morphology. The conformal welding idea can be extended from 2D shape space to 3D shape space, by combining conformal modules and curve diffeomorphism signatures to form 3D conformal welding signatures [29]. It describes the shape and location of 3D simple and nonintersecting contours on the embedding surface, and the underlying geometric correlation between each patches. It has been successfully used for classifying the human brain cortical surfaces among the healthy control group and the Alzheimer's Disease (AD) group. The structural MRI data were obtained from the Alzheimer's Disease Neuroimaging Initiative (ADNI) database (www.adni.loni.ucla.edu). In the computation, the brain cortical surface is partitioned into different functional regions, each region is conformally mapped to a canonical circle domain. This can be accomplished by using the Euclidean discrete Ricci flow method [9, 18, 28] as described in Algorithms 5.11 and 5.12. Then the boundary of each region induces a diffeomorphism from the unit circle to itself. The shapes of canonical spaces and the automorphisms of the unit circle form the signature as shown in Fig. 5.10. Theoretically, the signature is guaranteed to be a complete and global shape descriptor based on Teichmüller space theory and conformal welding theory. This method focuses on both local geometries of

---

Algorithm 5.11 Conformal module for doubly connected annuli

Input: A triangular mesh M, which is a doubly connected annulus.
Output: Conformal module of M.
 1: Set target curvature equal to zero everywhere.
 2: Compute a flat metric using Euclidean Ricci flow in Algorithm 5.2.
 3: Compute the shortest path τ between two boundaries.
 4: Slice the M along τ to get $\bar{M}$.
 5: Flatten $\bar{M}$ onto the plane using Euclidean embedding in Algorithm 5.4.
 6: Scale and translate the planar image of $\bar{M}$, such that the image of the outer boundary is aligned with the imaginary axis, the length of outer boundary is 2π.
 7: Use the exponential map $z \to e^z$ to map the planar image of $\bar{M}$ to an annulus.
 8: Measure the radii of inner and outer boundary circles r and R, respectively. The conformal module is given by $\frac{1}{2\pi} \ln \frac{R}{r}$.

Algorithm 5.12 Conformal module for triply connected annuli by Euclidean Ricci flow (generalized Koebe's method)

Input: A triangular mesh M, which is a triply connected annulus, with boundaries $\partial M = \gamma_0 - \gamma_1 - \gamma_2$.
Output: The Euclidean conformal module of M.
 1: **repeat**
 2:    Fill γ_1 by a topological disk, to get M_1.
 3:    Use Algorithm 5.11 to map M_1 to a canonical planar annulus $\tilde{M}_1$.
 4:    On $\tilde{M}_1$, remove the hole of γ_1 and fill the hole γ_2 to get M_2.
 5:    Use Algorithm 5.11 to map M_2 to a canonical planar annulus $\tilde{M}_2$.
 6:    On $\tilde{M}_2$, remove the hole of γ_2, to get $\tilde{M}$.
 7:    $M \leftarrow \tilde{M}$.
 8: **until** The curvatures on the boundaries are close to constants
 9: Suppose the center of γ_1 is z_0. Define a Möbius transformation $z \to \frac{z-z_0}{1-\bar{z}_0 z}$ to map the center of γ_1 to the origin.
10: Rotate the planar image, such that the center of γ_2 is on the imaginary axis.

---

functional regions and global geometric relations among them. The local-global view based on conformal geometry would be highly advantageous for biomarker research in AD.

5.3.2   *Teichmüller Space*

According to Theorem 3.1, all surfaces in real world are Riemann surfaces, therefore have conformal structures. All metric surfaces in real world can be classified by conformal equivalence. The space of all conformal classes is the Teichmüller space. The Teichmüller space coordinates are given by conformal modules, which are the complete invariants of conformal structures and intrinsic to surfaces themselves. The Teichmüller space coordinates can be applied for shape indexing directly.

The representations for the conformal module of a surface are not unique. One representation is based on Euclidean Ricci flow. The other is based on hyperbolic

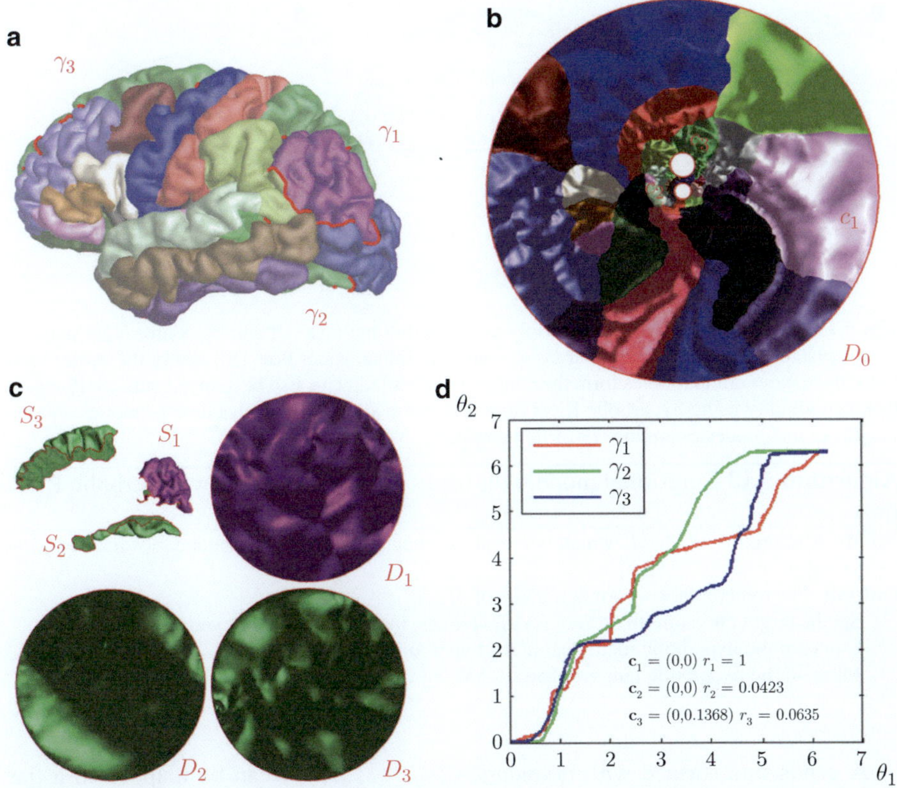

Fig. 5.10 Conformal welding signature via uniformization mapping for a hemisphere brain surface with 3 simple closed contours. The curve γ_i (**a**) surrounds the patch S_i (**c**) and is mapped to the boundary c_i of the circle domain D_i (**c**). γ_i is also mapped to the boundary of the base circle domain D_0 (**b**). The curves (**d**) demonstrate the diffeomorphisms for the three contours

Ricci flow. A topological multiply connected annulus, i.e., a genus zero surface with multiply boundaries, can be mapped to the canonical circle domain using Euclidean Ricci flow. Two circle domains are conformally equivalent if and only if they differ by a Möbius transformation. Suppose the boundaries of a circle domain D are $\partial D = C_0 - C_1 - C_2 - \cdots - C_n$, where C_k is a circle with the center $\mathbf{c}_k = (x_k, y_k)$ and the radius r_k, denoted as $(\mathbf{c}_k, r_k)$. One can normalize D by a Möbius transformation, such that C_0 becomes the unit circle, $\mathbf{c}_1$ is at the origin, and $\mathbf{c}_2$ is on the imaginary axis. Then the normalized circle domain is determined by $\mathrm{Mod}(D) = \{r_1, y_2, r_2\} \cup \{x_k, y_k, r_k, k > 2\}$, which is called the *conformal module* of the circle domain. Algorithms 5.11 and 5.12 explain the details for computing conformal module for doubly and triply connected annuli using Euclidean Ricci flow and the generalized Koebe's method [30, 31]. The proof for the convergence of the Koebe's algorithm can be found in [5].

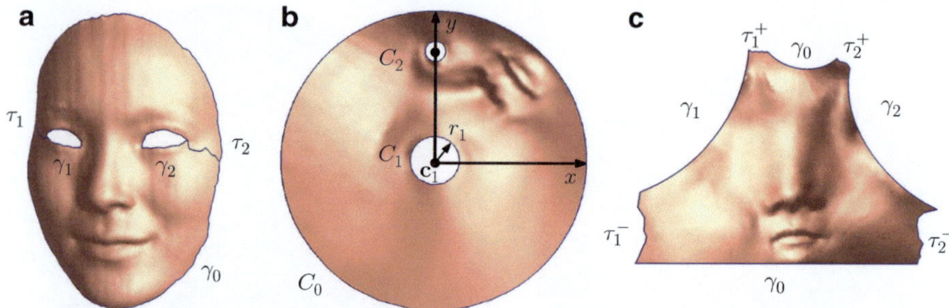

Fig. 5.11 Conformal modules for a triply connected domain (**a**). It can be conformally mapped onto a planar disk with two circular holes using Euclidean Ricci flow (**b**), where the centers and radii of the inner circular holes form the conformal module. It can also be mapped onto the Poincaré disk periodically using hyperbolic Ricci flow (**c**), where all boundaries become geodesics and the lengths of the geodesics are the conformal module

---

Algorithm 5.13 Conformal module for triply connected annuli by hyperbolic Ricci flow

---

Input: A triangular mesh M, which is a triply connected annulus with boundaries $\partial M = \gamma_0 - \gamma_1 - \gamma_2$.

Output: The hyperbolic conformal module of M.

  1: Set the target curvature to be zero for all vertices, including boundary vertices.
  2: Compute the hyperbolic metric using the hyperbolic Ricci flow.
  3: Measure the hyperbolic lengths of γ_0, γ_1 and γ_2, which give us the conformal module.

A genus one surface with boundaries $\{\gamma_0, \gamma_1, \ldots, \gamma_n\}$ can be mapped to a flat torus $\mathbb{E}^2/\Lambda$ with circular holes $\{(\mathbf{c}_k, r_k), 0 \le k \le n\}$, where Λ is a lattice transformation group $\{z \to z + m + n\omega, m, n \in \mathbb{Z}\}$, $\omega \in \mathbb{C}$. Then the conformal module is given by ω and $\{(\mathbf{c}_k, r_k)\}$.

As explained in Sect. 3.3.8, a compact high genus surface with boundaries can be conformally mapped to a hyperbolic surface $\mathbb{H}^2/\Lambda$ with circular holes $\{(\mathbf{c}_k, r_k)\}$, where Λ is the Fuchs group of the surface. Then the generators of Fuchs group and the circle centers and radii form the conformal module of the surface. Details can be found in [31].

Another way to define the conformal module for a surface with negative Euler number is as follows. We compute the uniformization hyperbolic metric, such that all the boundaries become geodesics. Then perform pants decomposition, such that all cuts are geodesic loops. The surface is glued by these pairs of pants. When two pairs of pants are glued along their common boundary, a twisting angle is introduced. The lengths of these cuts and the twisting angles are the conformal module. The twisting angles can be computed by the lengths of some geodesics in a specific set of homotopy classes. Each homotopy class corresponds to a Fuchsian transformation. The axis of the Fuchsian transformation is the unique geodesic in the homotopy class. Details can be found in [10]. Figure 5.11 illustrates two representations of the conformal module of a triply connected annulus. Algorithm 5.13 explains the details for computing the conformal module based on hyperbolic Ricci flow.

Application

Human Facial Surface Indexing. We capture 3D faces of different persons with the same expression. Then we extract the contours of the mouth and both eyes and remove the interior regions, making the surfaces into multiply connected domains, i.e., genus zero surfaces with four boundaries. We apply Euclidean Ricci flow to map the surface onto the unit disk with circular holes. Let the centers of the circular holes be $\mathbf{c}_k=(x_k, y_k)$, $k = 0, 1, 2, 3$ and the radii be r_k. We use a Möbius transformation to normalize the mapping such that the exterior circle becomes a unit circle, the first inner circle is centered at the origin, and the center of the second circle is on the positive side of the imaginary axis, i.e., $r_0 = 1$, $(x_0, y_0) = (0, 0)$, $(x_1, y_1) = (0, 0)$, and $x_2 = 0$, similar to the normalization in Fig. 5.11**b**. Then the conformal module is given by $(r_1, y_2, r_2, x_3, y_3, r_3)$. The Teichmüller space is 6 dimensional. Conformal modules can be used for 3D human face shape indexing. Details can be found in [28].

Vestibular System Morphometry. Adolescent Idiopathic Scoliosis (AIS) characterized by 3D spine deformity affects about 4% schoolchildren worldwide. One of the prominent theories of the etiopathogenesis of AIS was proposed to be the poor postural balance control due to the impaired vestibular function. Thus, the morphometry of the vestibular system (VS) is of great importance for studying AIS. The VS is a genus three structure situated in the inner ear and consists of three semicircular canals lying perpendicular to each other. The high genus topology of the surface poses great challenge for shape analysis. An effective method is to analyze shapes of high genus surfaces by considering their geodesic spectra. The key is to compute the canonical hyperbolic geodesic loops of the surface, using hyperbolic Ricci flow, as shown in Fig. 5.12. The geodesic spectra effectively measure shape differences between high genus surfaces up to hyperbolic isometry. Experimental results on the VS of normal and AIS subjects show the effectiveness of the algorithm and reveal the statistical shape difference between two groups. Details can be found in [24].

Brain Mapping. The human brain cortical surfaces can be reconstructed from CT images and discretized as triangular meshes. The brain cortical surface, a topological sphere, can be mapped to a unit sphere, as shown in Fig. 3.4. All the geometric details are exposed onto the sphere. The morphometry analysis can be constructed on the canonical domain. Besides that, by using the landmarks on the brain, the brain surface is sliced open with holes, where each landmark generates a hole. Therefore, the brain cortical surface with holes can be either mapped to a circle domain where each hole is mapped to a circle, or a hyperbolic convex domain where each hole is mapped to a geodesic in hyperbolic space. The conformal modules can be extracted for brain surface shape analysis purposes. Figure 5.13 shows an example of hyperbolic mapping for a brain cortical surface with one boundary and three interior landmarks.

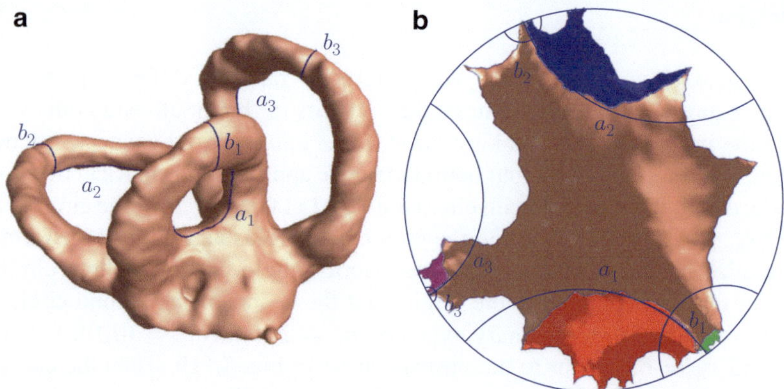

Fig. 5.12 Computing the geodesic loops for each homotopy class of a VS surface. All the geodesic loops $\{a_i, b_i\}, i = 1, 2, 3$ (**a**) form the homotopy group basis on the Poincaré disk (**b**). The lengths of the geodesic loops in $\mathbb{R}^3$ form the geodesic spectra

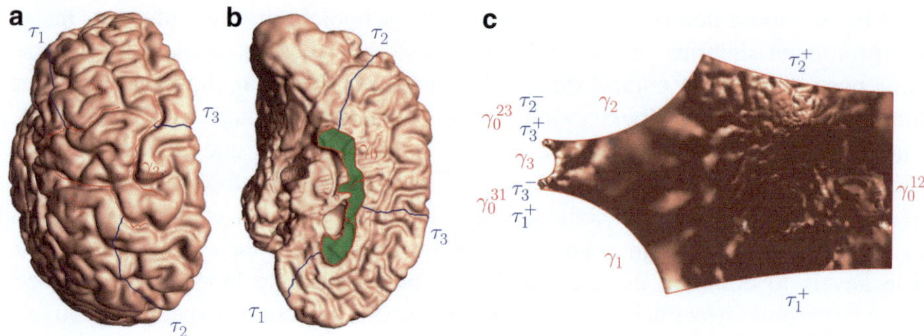

Fig. 5.13 Brain mapping by hyperbolic Ricci flow. A human brain surface with one boundary γ_0 (**b**) is sliced open by three interior landmarks $\gamma_k, k = 1, 2, 3$ (**a**), then it becomes a genus zero surface with four boundaries. By hyperbolic Ricci flow, it is mapped to a hyperbolic convex polygon on Poincaré disk (**c**) where the domain is sliced open by three curves τ_k (**b**) between γ_0 and γ_k, respectively

It is crucial to register different brain cortical surfaces in brain imaging field. Since brain cortical surfaces are highly convoluted and different people have different anatomic structures, it is quite challenging to find a good registration (or mapping) between them. By finding an automorphism of the canonical shapes, the registration between surfaces can be easily established. The computation of the spherical, Euclidean, and hyperbolic brain mappings is based on the discrete Ricci flow method, as described in Algorithm 5.2. The details of brain surface mapping and its applications can be found in [3, 18].

Human Facial Surface Dynamics Analysis. Suppose a sequence of facial surfaces with different expression changes are captured. By computing the time-dependent

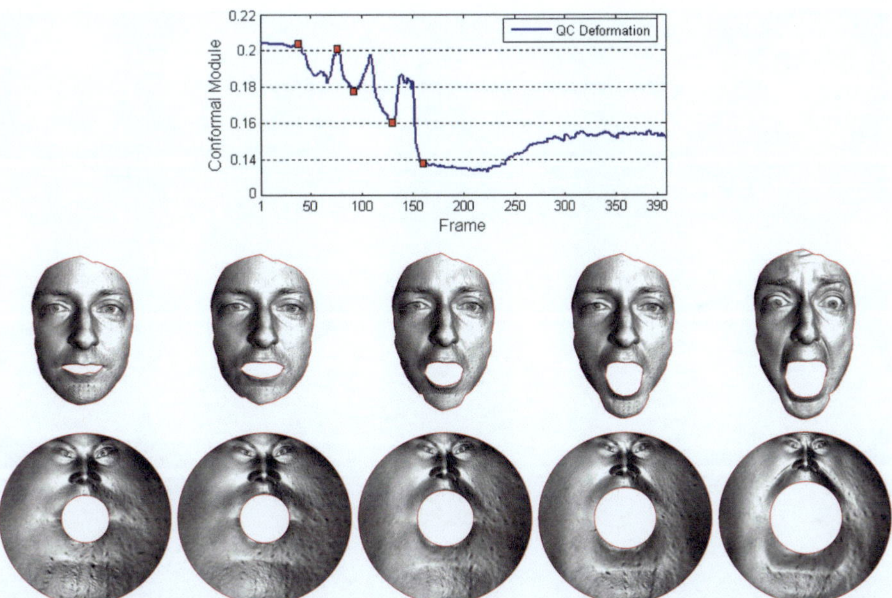

Fig. 5.14 Conformal modules of a sequence of human facial surfaces with dynamic expressions. All the faces are mapped to a unit disk with one concentric circle with radius r. The conformal module is defined as $\frac{1}{2\pi}\ln\frac{1}{r}$, which decreases monotonically as the inner circle radius increases

conformal module $R(t)$, we can analyze the expression. Figure 5.14 shows such an example. A sequence of male facial surfaces from calm expression to frightened expression is acquired using 3D camera system [17]. We remove the mouth region and compute the conformal module. It is easy to see that the human facial expression intensity is proportional to the norm of the conformal module. Details can be found in [20].

5.3.3 Surface Conformal Representation

According to Theorem 3.3, a surface embedded in $\mathbb{R}^3$ is fully determined by the conformal factor function λ and the mean curvature function H defined on the conformal structure R, then we use (R, λ, H) as the representation of the surface. R gives the Riemann surface structure, (R, λ) gives the Riemannian metric, and (R, λ, H) gives the embedding of the surface in $\mathbb{R}^3$. The Riemann surface R can be represented using its conformal module.

Given a surface S embedded in $\mathbb{R}^3$, it has the induced Euclidean metric $\mathbf{g}$. We compute its uniformization metric $\mathbf{g}_0$ using Ricci flow and obtain the Riemann surface structure R and the conformal factor $\mathbf{g} = e^{2\lambda}\mathbf{g}_0$. We choose local conformal parameters z of R. The mean curvature is computed as

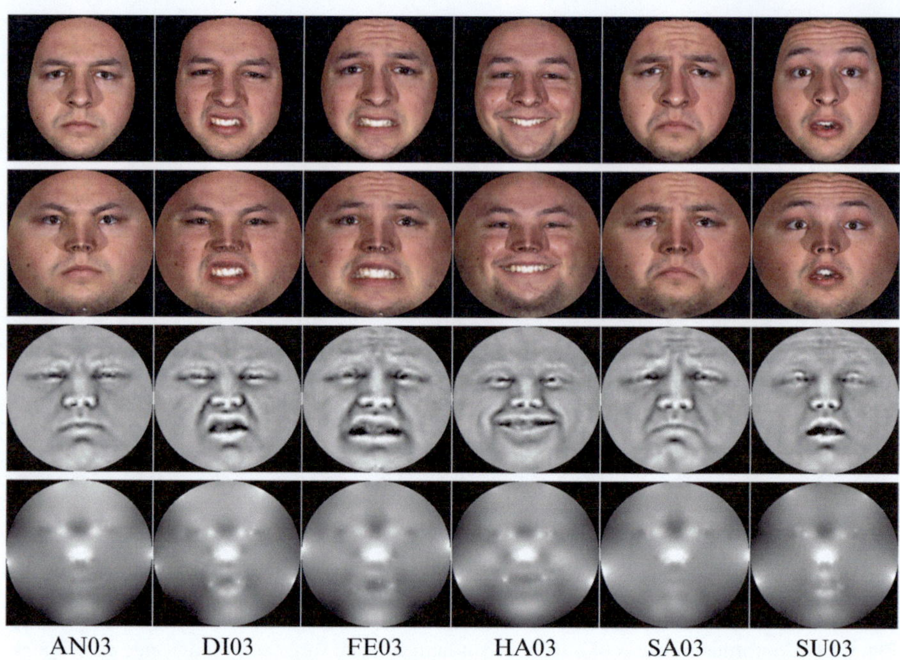

AN03   DI03   FE03   HA03   SA03   SU03

Fig. 5.15 Conformal representations for 3D human facial surfaces with expressions with the 3rd intensity level. The subject is M0021 in BU3D database. From *upper* to *bottom*: human face images in camera view, conformal maps with human facial textures, mean curvature images, and conformal factor images for 6 expressions, respectively

$$H(z) = \left\langle \frac{4}{e^{2\lambda(z)}} \frac{\partial^2 \mathbf{r}(z)}{\partial z \partial \bar{z}}, \mathbf{n}(z) \right\rangle,$$

where $\mathbf{r}(z)$, $\mathbf{n}(z)$ are the position vector and the normal of the surface, respectively. Figure 5.15 illustrates the conformal representations of a 3D human facial surface with 6 different expressions. Each column gives the conformal representation for a different expression. From *upper* to *bottom*, the *first* row shows the original surface with texture, the *second* row is the image of conformal mapping, and the *third* and the *fourth* rows show the conformal factor and the mean curvature functions defined on the conformal parameter domain, respectively, which are color encoded as gray-level images. Details can be found in [23].

Measuring the distance among shapes has fundamental importance in practice. In the following, we give one way to define shape distance between two surfaces based on their conformal representations and the diffeomorphic registration between them. Given two surfaces $(S_1, \mathbf{g}_1)$ and $(S_2, \mathbf{g}_2)$, we compute their conformal representations using Ricci flow, (R_1, λ_1, H_1) and (R_2, λ_2, H_2). Then we can define the distance between them as

$$d(S_1, S_2) := \min_{f:R_1 \to R_2} \left[\int_{R_1} (\lambda_1 - \lambda_2 \circ f)^2 \right]^{\frac{1}{2}} + \left[\int_{R_1} (H_1 - H_2 \circ f)^2 \right]^{\frac{1}{2}},$$

where we restrict the mapping f to be a diffeomorphism. It is straightforward to show that $d(\cdot, \cdot)$ satisfies the triangle inequality.

The optimal diffeomorphism $f : R_1 \to R_2$ can be obtained using quasi-conformal geometric method introduced in Sect. 5.2.3. If we emphasize the efficiency, then the mapping f could be chosen to be harmonic mappings. If the surfaces are with negative Euler number, then we can equip the target with the hyperbolic metric, under which the harmonic mapping is unique in each homotopy class and can be computed using the method introduced in Sect. 5.2.2.

Application

3D Human Facial Expression Recognition. Surface conformal representation can be applied for 3D human facial expression recognition. An automatic 3D expression recognition system has been built up and tested on the BU3D database [19]. For each subject, the database includes one neutral and six prototypical non-neutral expressions, such as happiness (HA), sadness (SA), anger (AN), fear (FE), surprise (SU), and disgust (DI), with four intensity levels. As shown in Fig. 5.15, the conformal representations for all facial surfaces are automatically computed and classified using standard classifiers. The preliminary experimental results show the current approach is highly competitive [23].

Symmetry Detection. Symmetry analysis is an important task in shape analysis. There are intrinsic symmetry and extrinsic symmetry. Suppose $(S, \mathbf{g})$ is a metric surface. A mapping $f : S \to S$ is isometric if the pullback metric equals the original metric, $\mathbf{g} = f^* \mathbf{g}$. All the isometric automorphisms of S form a group, which is the *intrinsic symmetry group* of the surface. Furthermore, if an isometric automorphism f also preserves mean curvature, namely, $H \circ f = H$, then f is also an extrinsic symmetry.

Ricci flow can help detect such type of symmetry. The key observation is that Ricci flow preserves intrinsic symmetry. Suppose $\mathbf{g}(t)$ is the Riemannian metric during the flow at time t. If an automorphism $f : S \to S$ is isometric at the beginning, $f^* \mathbf{g}(0) = \mathbf{g}(0)$, then for all time t, $f^* \mathbf{g}(t) = \mathbf{g}(t)$. Let $t \to \infty$, then $\mathbf{g}(\infty)$ is one of the three canonical metrics, spherical, Euclidean, or hyperbolic metric. Then the intrinsic symmetry group must be a subgroup of the rigid motion group under the canonical metric.

Furthermore, the symmetry axis or the fixed point sets of the isometric mapping f, $\{p \in S | f(p) = p\}$, must be geodesics. The geodesics under canonical metrics are spherical, Euclidean, or hyperbolic lines. Therefore, by using Ricci flow, intrinsic symmetry detection can be carried out in a much easier way. Figure 5.16 demonstrates the symmetry on both the original surface and the conformal parameter domain obtained by the Ricci flow method.

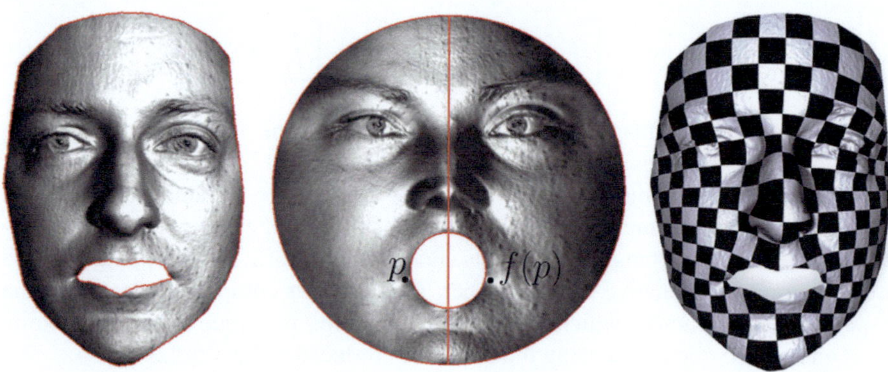

Fig. 5.16 Symmetry detection by Ricci flow

Fig. 5.17 Symmetric conformal mapping for a human facial surface using Euclidean Yamabe flow. The property of symmetry preserving is illustrated in the flat image and the checkerboard texture mapping image, where the conformality and the area distortion are also visualized

In addition, intrinsic symmetry can be incorporated into Ricci flow to generate *symmetric conformal mapping*, preserving the symmetry in canonical domain, as shown in Fig. 5.17. The involvement of symmetry can improve the robustness when registering the occluded surfaces with noisy boundaries [22].

References

1. Center, M.M., Jemal, A., Smith, R.A., Ward, E.: Worldwide variations in colorectal cancer. CA: A Cancer J. Clin. **59**(6), 366–378 (2009)
2. Gu, X., Luo, F., Sun, J., Yau, S.T.: Variational principles for Minkowski type problems, discrete optimal transport, and discrete Monge-Ampere equations. arXiv:1302.5472 pp. 1–13 (2013)
3. Gu, X., Wang, Y., Chan, T.F., Thompson, P.M., Yau, S.T.: Genus zero surface conformal mapping and its application to brain surface mapping. IEEE Trans. Med. Imag. **23**(7), 949–958 (2004)

4. Haber, E., Metaxas, D.N., Axel, L.: Motion analysis of the right ventricle from MRI images. In: International Conference on Medical Image Computing and Computer Assisted Intervention (MICCAI'98), Springer-Verlag, Berlin, Heidelberg, pp. 177–188 (1998)
5. Henrici, P.: Applied and Computational Complex Analysis, vol. 3. Wiley-Intersecience, New York (1988)
6. Hernandez, C., Vogiatzis, G., Brostow, G.J., Stenger, B., Cipolla, R.: Non-rigid photometric stereo with colored lights. In: IEEE International Conference on Computer Vision (ICCV'07), Institute of Electrical and Electronics Engineers (IEEE), pp. 1–8 (2007)
7. Hong, L., Muraki, S., Kaufman, A., Bartz, D., He, T.: Virtual Voyage: Interactive Navigation in the Human Colon. ACM SIGGRAPH, New York, pp. 27–34 (1997)
8. Horner, M.J., Ries, L.A.G., Krapcho, M., Neyman, N., Aminou, R., Howlader, N., Altekruse, S.F., Feuer, E.J., Huang, L., Mariotto, A., Miller, B.A., Lewis, D.R., Eisner, M.P., Stinchcomb, D.G. In: Edwards, B.K. (eds.) SEER Cancer Statistics Review, 1975–2006, National Cancer Institute, Bethesda, MD (2009)
9. Jin, M., Kim, J., Luo, F., Gu, X.: Discrete surface Ricci flow. IEEE Trans. Visual. Comput. Graph. **14**(5), 1030–1043 (2008)
10. Jin, M., Zeng, W., Ding, N., Gu, X.: Computing Fenchel-Nielsen coordinates in Teichmüller shape space. Int. J. Commun. Inform. Syst. **9**(2), 213–234 (2009)
11. Johnson, C.D., Dachman, A.H.: CT colography: the next colon screening examination. Radiology **216**(2), 331–341 (2000)
12. Lui, L.M., Wong, T.W., Zeng, W., Gu, X., Thompson, P.M., Chan, T.F., Yau, S.T.: Detecting shape deformations using Yamabe flow and Beltrami coefficients. J. Inverse Prob. Imag. **4**(2), 311–333 (2010)
13. Lui, L.M., Wong, T.W., Zeng, W., Gu, X., Thompson, P.M., Chan, T.F., Yau, S.T.: Optimization of surface registrations using Beltrami holomorphic flow. J. Scient. Comput. **50**(3), 557–585 (2012)
14. Lui, L.M., Zeng, W., Chan, T., Yau, S.T., Gu, X.: Shape representation of planar objects with arbitrary topologies using conformal geometry. In: European Conference on Computer Vision (ECCV'10), Crete, Greece (2010)
15. Qiu, F., Fan, Z., Yin, X., Kauffman, A., Gu, X.: Colon flattening with discrete Ricci flow. In: International Conference on Medical Image Computing and Computer Assisted Intervention (MICCAI'09), London, England (2009)
16. Sharon, E., Mumford, D.: 2D-shape analysis using conformal mapping. Int. J. Comput. Vision **70**, 55–75 (2006)
17. Wang, Y., Gupta, M., Zhang, S., Wang, S., Gu, X., Samaras, D., Huang, P.: High resolution tracking of non-rigid motion of densely sampled 3d data using harmonic maps. Int. J. Comput. Vision **76**(3), 283–300 (2008)
18. Wang, Y., Shi, J., Yin, X., Gu, X., Chan, T.F., Yau, S.T., Toga, A.W., Thompson, P.M.: Brain surface conformal parameterization with the Ricci flow. IEEE Trans. Med. Imag. **31**(2), 251–264 (2012)
19. Yin, L., Wei, X., Sun, Y., Wang, J., Rosato, M.J.: A 3D facial expression database for facial behavior research. In: IEEE International Conference on Automatic Face and Gesture Recognition (FG'06), vol. 0, pp. 211–216. IEEE Computer Society, Los Alamitos (2006)
20. Zeng, W., Gu, X.: 3D dynamics analysis in Teichmüller space. In: ICCV Workshop on Dynamic Shape Capture and Analysis Recognition (4DMOD'11), Barcelona, Spain (2011)
21. Zeng, W., Gu, X.: Registration for 3D surfaces with large deformations using quasi-conformal curvature flow. In: IEEE Conference on Computer Vision and Pattern Recognition (CVPR'11). Colorado Springs, Colorado (2011)
22. Zeng, W., Hua, J., Gu, X.: Symmetric conformal mapping for surface matching and registration. Int. J. CAD/CAM **9**(1), 103–109 (2009)
23. Zeng, W., Li, H., Chen, L., Morvan, J.M., Gu, X.: An automatic 3D expression recognition framework based on sparse representation of conformal images. In: IEEE International Conference on Automatic Face and Gesture Recognition (FG'13). IEEE Computer Society, Shanghai (2013)

24. Zeng, W., Lui, L., Shi, L., Wang, D., Chu, W.C., Cheng, J., Hua, J., Yau, S.T., Gu, X.: Shape analysis of vestibular systems in adolescent idiopathic scoliosis using geodesic spectra. In: International Conference on Medical Image Computing and Computer Assisted Intervention (MICCAI'10), Beijing, China (2010)
25. Zeng, W., Lui, L.M., Luo, F., Chan, T., Yau, S.T., Gu, X.: Computing quasiconformal maps using an auxiliary metric with discrete curvature flow. Numeriche Mathematica **121**(4), 671–703 (2012)
26. Zeng, W., Luo, F., Yau, S.T., Gu, X.: Surface quasi-conformal mapping by solving Beltrami equations. In: IMA International Conference on Mathematics of Surfaces XIII, Springer-Verlag, Berlin, Heidelberg, pp. 391–408 (2009)
27. Zeng, W., Marino, J., Gurijala, K., Gu, X., Kaufman, A.: Supine and prone colon registration using quasi-conformal mapping. IEEE Trans. Visualiz. Comput. Graph. **16**(6), 1348–1357 (2010)
28. Zeng, W., Samaras, D., Gu, X.D.: Ricci flow for 3D shape analysis. IEEE Trans. Pattern Anal. Mach. Intell. **32**(4), 662–677 (2010)
29. Zeng, W., Shi, R., Wang, Y., Yau, S.T., Gu, X.: Teichmüller shape descriptor and its application to Alzheimer's disease study. Int. J. Comput Vision. **105**(2), 155–170 (2013)
30. Zeng, W., Yin, X., Zhang, M., Luo, F., Gu, X.: Generalized Koebe's method for conformal mapping multiply connected domains. In: SIAM/ACM Joint Conference on Geometric and Physical Modeling (SPM'09), San Francisco, California (2009)
31. Zhang, M., Li, Y., Zeng, W., Yau, S.T., Gu, X.: Canonical conformal mapping for high genus surfaces with boundaries. Int. J. Comput. & Graph. **36**(5), 417–426 (2012)

Index

W. Zeng and X.D. Gu, *Ricci Flow for Shape Analysis and Surface Registration*,
SpringerBriefs in Mathematics, DOI 10.1007/978-1-4614-8781-4,
© Wei Zeng, Xianfeng David Gu 2013